T0191857

Lecture Notes in Earth System Sciences

More information about this series at http://www.springer.com/series/10529

Gabriele Morra

Pythonic Geodynamics

Implementations for Fast Computing

Gabriele Morra
Department of Physics and School
of Geoscience
University of Louisiana at Lafayette
Lafayette, LA
USA

ISSN 2193-8571 ISSN 2193-858X (electronic)
Lecture Notes in Earth System Sciences
ISBN 978-3-319-85725-1 ISBN 978-3-319-55682-6 (eBook)
DOI 10.1007/978-3-319-55682-6

Printed on acid-free paper

This Springer imprint is published by Springer Nature
The registered company is Springer International Publishing AG
The registered company address is: Gewerbestrasse 11, 6330 Cham, Switzerland

Foreword

In recent years, there arises a great need for a freshly crafted beginners text book in computational geosciences, because of the rising levels of capable university students with improved background in mathematics and physical sciences. Today in the USA and other countries, many freshmen and high school seniors have sufficient background and sophistication from advanced placement courses to appreciate a firm knowledge of computational knowledge is needed to solve real problems in geosciences without going through the drudgery of analytical solutions and analysis typically found in traditional textbooks, written in the twentieth century. So, they are more than ready to try some new pedagogical undertaking, as present in this volume prepared by Gabriele Morra.

The most direct link between simple physical laws and the real world is the numerical solutions of time-dependent partial differential equations from real systems in terms of realistic examples with physical meaning. On the one hand, experimental technique and data collection have seen such dramatic progress that by now many fundamental properties of geophysical fluid dynamics can be demonstrated in laboratory experiments, such as thermal convection and mixing. On the other hand, great efforts are being made to exploit ideas from nonlinear dynamics in cases where the system is not necessarily deterministic but the data displays more structure than can be captured by traditional analytical methods. Problems of this kind are typical in many areas of science where multi-scale phenomena prevail as in geosciences but in many other sciences, like in biology and physiology.

This book by Morra provides the young students in geosciences with a sound knowledge of tools such as Python and modern numerical techniques for more incursions into other burgeoning areas such as Big Data and High-Performance Computing. I hope that this book will inspire others to write similar books for future generation of students. This book will indeed bring a fresh approach to computational geosciences.

February 2016

David A. Yuen
University of Minnesota
Minneapolis, MN
USA

Preface

Adults, like children, learn by playing. With this insight in mind, I wrote this book as a non-exhaustive and non-overambitius text that aims at introducing computational geodynamics to young students, either undergraduates or beginning graduates. The goal, more than trying to cover a topic that is way too vast for a book to contain it, is to show some of the fundamental strategies available with a strong focus on practical, playful, easy-to-use techniques.

Python is now a standard tool for programming in science and even more in the industry, where the explosion of the Big Data, mining, and artificial intelligence research fields has convinced many professionals to learn this simple and powerful language. While I write, Python, combined with its visualization and numerical libraries, is replacing MATLAB in many scientific and technological contexts, because it offers the same capabilities but with the great advantage of being open source. Open projects attract more developers which, in the long term, always overcomes closed/commercial tools if the user base is sufficiently large. Many developers are in this moment turning existing C and Fortran libraries into Python, among them the ones that allow parallel computing, programming NVIDIA GPUs, running machine learning algorithms, and many more. Although Numerical Python is not the only fast growing scientific/technological language, as others such as Ruby and Scala are emerging as well, Python is the present state of the art in scientific computing.

Geodynamics, the study of the evolution of our planet and similarly of the other solid planets, is a fast evolving field. It constantly englobes new research topics ranging from the dynamics of the Earth's core, to the convection of its slowly deforming solid mantle, from the study of the dynamics of the Geoid (the *gravitational* shape of the Earth) to the global resonance of the entire planet to the stimulus of a large Earthquake on its surface, from the dynamics of the two-phase gas-magma system in a volcano conduit to the estimation of the risks related to small and giant eruptions, from the dramatic formation of our planet to the increase of its mineralogical variety during its evolution, from the detection of the chemical spectrum of the atmosphere of some exoplanets modelling their interior dynamics.

Given that most of these systems are at the same time complex and inaccessible to direct measurements, numerical modeling plays a key a role to their understanding. I have been teaching numerical modeling applied to geosciences, and geodynamics in particular, in my past academic positions at the University of Sydney, Seoul National University, and presently at the University of Louisiana at Lafayette. In my courses I have used numerous books that covered several aspects of geodynamics, however many of the key progresses in the past 30 years have been driven or confirmed by well crafted numerical experiments. I have, therefore, decided to write a textbook that focuses on the practice of designing these numerical experiments, organized as a hands-on manual, almost like one of these online tutorial that programmers use every day, but with a greater structure, designed to teach undergraduate and graduate students to create their own numerical models.

In my courses, I verified that the insight that students gain by designing and writing a code is often greater than the understanding gained by the solution itself. To create a software that simulates the Earth's interior behavior is like to teach geodynamics to a machine, and requires a very clear understanding of the underlying physics behind geological phenomena. This book combines the material taught in three different occasions. One is the introduction to computation to honor sophomore students of general physics at the University of Louisiana at Lafayette. The second is a course of computational geophysics taught at the same university to graduate students during the academic years 2013/2014 and 2014/2015. Finally, I added advanced topics such as the one taught at a summer school on large-scale Boundary Element Method (BEM) at the University of Brescia, Italy, in 2011.

The book is structured in four parts. In the first one, I introduce in three chapters a (i) bird's eye view of the computational capabilities of Python, (ii) the visualization tools available in Python, and (iii) how to use the powerful Numerical Python libraries and other numerical tools to embed C, OpenMP, MPI capabilities to our Python programs.

In the second part, I illustrate few examples of how to use Numerical Python to solve typical problems of a standard general physics course. These problems are normally solved only using analytical tools, while here I illustrate how to approach them numerically, and at the same time familiarize with momentum and energy equations, the main ones that are solved in geodynamic modeling. The goal of this part is to introduce the less expert reader to using Python, solve numerically some simple problems, and teach at the same time how to calculate standard physics quantities such as momentum, work, power, dissipative energy, and the action. In this part, we will for the first time learn how to numerically calculate derivatives and integrals, how to monitor the accuracy of a numerical solution, and how to visualize the results of a computational model.

In the third part of the book, I introduce the main laws of continuum mechanics that every geodynamicist needs to know, diffusion and momentum equations in a continuum context, and how to use Finite Differences, and the advanced *Particle in Cell (PIC)* technique to solve them. The goal of this part is to describe how to write compact and elegant but still very fast routines that allow implementing a sophisticate and advanced code such as *PIC* in simple and straightforward manner. Applications to solving the equations that describe strongly viscous flow, porous media flow, elasticity, both in a linear and nonlinear setup are introduced in a simplified and introductory manner. Also some important issues such as numerical stability/instability are shown using examples.

The fourth and last part of the book covers a small set of more advanced topics such as numerical approaches that do not require the discretization of the space, but are built from the summation of fundamental solutions. This will give the reader the opportunity to familiarize with tree-based codes, that are at the core of many modern numerical techniques and can efficiently solve the many body interaction problem that applies to many aspects of geosciences and beyond, such as planetary science. A specific application to the calculation of the dynamics of a multiphase fluid is finally introduced and applied to the motion of gas bubbles in a magmatic system.

I believe that this book still lacks much material that I would like to add. In particular more details on the solution techniques for the non-linear Navier-Stokes equation, lagrangian multiplier method, detailed implementations of the tree fast multipole method, and many more examples of how to parallelize and benchmark these codes. I have however to stop here and plan these additions for a second edition of this book. I hope that the very incomplete material of this first edition will anyway result useful to some young students who approach the fascinating field of computational geodynamics for the first time.

Who Should Read this Book

This book has been written for ambitious undergraduate college students and for graduate students in geophysics/geodynamics. I eliminated most of the calculus based derivations, while I focus on in implementations and examples, with the idea that it is the 'practice' that makes the 'scientist', and that 'creativity' is the product of 'perseverance'. This book is, therefore, designed for readers of every background who desire to learn about the behavior of our planet by explicitly modeling it.

This book is also addressed to more advanced practitioners who have been modelling geodynamics using different programming languages than Python and aim at trying new numerical techniques. In many ways Python is the best language for prototyping scientific implementations, and also for running some high performance simulations. I remember that when I begun my Ph.D. in Geodynamics, I dreamed about learning everything, mantle convection, earthquakes, seismic wave propagation, oceanography, glacial rebound, exoplanets, surface processes, and so on, but quickly I realized that to write an efficient, robust, and reliable software for modeling is a huge work. I certainly do not affirm that now that with the scientific environment of Python, and the emerging new tools that come from Machine Learning, this is today possible, but certainly presently a numerical modeler can aim at a much broader and multidisciplinary project than it was possible 15 years ago.

How Should this Book be Read

This volume is structured with increasing difficulty, starting from a level accessible to freshman college student. Through the book the level rises until reaching some topics chosen from the present-day geodynamics research. The goal of this book is to offer to the reader the key instruments not only to create general and powerful computational tools, but also a clear understanding of the difficulty of implementing them.

The reader is expected to test all the examples proposed and try to do as many exercises as possible. Real learning is achieved only by writing a software by our own, and this is much easier to achieve in Python compared to standard scientific languages such as C and Fortran.

The reader with little background in computational sciences will find easier to study the book in the same order as the material is presented, but the expert programmer can safely skip the first chapters where the main tools for achieving high performance with Python are introduced. Only from Chap. 7 the book starts to build on past shown examples, therefore sequential reading is recommended.

Great software for Geodynamics already exist today, such as *Access*, *Underworld*, and Terra. Still it is possible today, by using smart programming to quickly but gradually guide a student to building a geodynamic modeling software. The goal of this book is not to push the student toward competing against large projects, but to prepare them to understand how they work in order to work within these projects, or to develop new modules for running techniques never implemented elsewhere. In other word the idea of this book is to invite students to learn by experimenting in freedom.

Lafayette, Louisiana, USA Gabriele Morra
December 2016

Acknowledgements

This book was written for one year and a half from summer 2015 until the end of 2016. During this time I took a year leave from my university position to stay close to my family, who lived in another continent. Far from the university environment, from teaching and from the possibility to go to meetings or anyway interact with other scientists, I decided to write this book. I have, therefore, to thank the many people who supported and stimulated me, both emotionally and physically during this time.

Certainly in this time my family made the greatest sacrifice. Much of time that I planned to spend, during weekends and holidays, with my wife Dorothea and my daughters Marlen and Renee, was instead invested into writing many of the pages of this book. Many nights were spent into rewriting entire chapters of this book, with the result that I was not fully there the day after for them. I thank them immensely for their support and patience.

I want to thank my graduate student Prasanna Mahesh Gunawardana. Much of the material in this book has been shown to him first and he used to develop a software. Based on his feedback, and his questions on what to better explain and how, I added many of the routines that I selected for this book. I also want to thank you my entire class of the computational geological modeling at UL at Lafayette of 2014: with them I tested first how to teach Numerical Python to young graduate geology students.

I would like to thank those who provided me with useful feedback on the entire manuscript or on part of it. Among them, Sang-Mook Lee, Erik Sevre, Fabio Capitanio, Manuele Faccenda. I would like to thank Paul Russell Woodward for suggesting to use the word *Pythonic* in the title. A special thank to Peter Mora who carefully read the entire book and made many suggestions.

Last but not least, I have to thank David A. Yuen. His continuous support and patience has been incredible. Every time I started to become distracted with some new interesting topic he immediately called me back and forced me to focus into finishing this book. Without his tireless stimulus, this book would have probably never seen the light.

Contents

Part I
Introduction to Scientific Python

In the first chapter a birds's eye view on how to use Python for scientific purposes is proposed. Then a chapter is devoted to visualization of small and large scientific data and modeling results. Examples range from retrieving and plotting time series of Geodetic data obtaining and visualizing large,open geophysical datasets.

In the third chapter, Numerical Python (Numpy) functions are introduced focusing on fast manipulation of *ndarrays*. It is illustrated in detail how to operate with indexing, slicing, and Boolean operators on arrays in order to achieve the maximum performance, with specific applications to linear algebra. Finally, it is shown how nonexisting functions can be added using *Cython*, *Numba* and others. Parallel programming tools for Python are quickly introduced.

Chapter 1
Bird's Eye View

Readability counts.
Simple is better than complex.
Complex is better than complicated.
If the implementation is hard to explain, it's a bad idea.
If the implementation is easy to explain, it may be a good idea.
— from The Zen of Python of T. Peters

Abstract In this chapter it is reviewed the history of Python, the main differences between the 2.x and the 3.x versions, and how to choose the appropriate version to start with. It is then shown how to work either with Python in an interactive mode, using iPython (now Jupyter), or in the standard productive mode for large projects, with the main two implementations at present: Anaconda Spyder and Enthought Canopy. The main libraries used through the book are quickly introduced, among them MatPlotLib, NumPy, and Cython.

1.1 Bird's Eye View

The purpose of this book is to introduce the reader to some fundamental techniques used to model the evolution of solid planets, such as Earth, using computers of small as well as great power. I will treat these numerical methods from a very general standpoint, using examples of how to solve fundamental equations in physics and fluid dynamics, and only finally showing the application to geodynamics. The intention is to leave to the reader the freedom to develop the geodynamic model that they desire, without sticking to a specific numerical approach or predetermined assumptions for geodynamics.

To teach these techniques I will employ *Python*, a free programming language that uniquely shares the property of being simple, powerful, and widely used in science and engineering. During the development of Python many projects have grown together with it. For example, *NumPy* has been developed to increase performance for all vectorial operations, the main tool necessary to solve numerical problems. *MatPlotLib* has been introduced to offer a visualization experience similar to the

G. Morra, *Pythonic Geodynamics*, Lecture Notes in Earth System Sciences,
DOI 10.1007/978-3-319-55682-6_1

commercial package MATLAB, but has now grown to being the standard for Python users.

In the first part of the book I will teach to use *iPython* (interactive Python), a powerful front end that allows to easily familiarizing with the language and testing it. Later in the book, I will present more advanced and sophisticated implementations that can be run from a bash shell or from a comprehensive environment. While I write, the two largest projects aimed at developing a universal environment for Python are *Anaconda Spyder* (`https://www.continuum.io/`) and *Enthought Canopy* (`https://www.enthought.com/products/canopy/`). Both are available and come with the numerical and visualization libraries on Linux, OS X, and Windows platforms, allowing every user to skip the tedious procedure of installing a large number of library set.

1.2 History

The genesis of Python goes back to the work of the Dutch programmer Guido van Rossum, who created the first implementation in December 1989, and released it for the first time in 1991. For this reason, he has been informally nominated *Benevolent Dictator for Life* by the Python developers community. This simply means that in case of an unsolvable controversy he will take the definitive decision.

Since then, Python together with other popular dynamic programming languages such as Perl and Ruby, has taken off becoming the standard in many sectors of software development. Definitive versions have been released on January 1994 (1.0) and on October 2000 (2.0).

The development of Python faced a twist on December 2008 when a backwards-incompatible 3.0 version was released. Suddenly some Python software could run only on Python 2.x while others only on Python 3.x, which irritated many users. After 8 years from the first release of Python 3.x, however, the new software is almost exclusively developed for the new version of Python. In fact, Python 2.x will continue to be supported only until 2020. After that, Python 3.x will become the only supported version, therefore it is wise to write your software on Python 3.x only. It is important to emphasize that most incompatibilities between Python 2.x and 3.x are not related to the content of this book, but concern mostly I/O (input/output). Still some new features are essential in allowing Python programmers to use *threads* that are essential in the connected present world, therefore Python 3.x has to be the choice for the new developer. Most examples of this book run equally well on both platforms, with some possible conflicts arising when exporting the results and data.

In many ways computer science is an art. When two different implementations exist for solving the same problem, computer scientists debate about which one is the 'best' one, or the 'most elegant', often related to subjective opinions. Languages are, therefore, preferred by computer scientists when for each problem there is clearly one and only one 'best' way to implement its solution. Python often displays this property, however as T. Peters in *The Zen of Python* observed: "There should be one —and preferably only one —obvious way to do it. Although that way may not be obvious at first unless you're Dutch."

Many books can introduce you to the Python Language. For the ones who did not have any experience at all with programming, ***Think Python*** from Allen B. Downey, O'Reilly, is a clear and simple guide. The book is presently freely available from the website `http://www.greenteapress.com/thinkpython/`, where a PDF copy of the book can be downloaded, both for the 2.x and 3.x versions. More details, the complete manuals, and tutorials are freely available on the Python Website (`http://python.org/`), as well as many on other sources such as Code Academy (`https://www.codecademy.com/learn/python`) and Python Course (`http://www.python-course.eu/`).

The rise of Python has triggered the development of a huge number of tools to expand its functionality. Its ability to integrate existing compiled software, C and C++ above all, but also Fortran, has allowed building hybrid projects, where Python plays the role of glue between them. In the past years, in particular, Cython (`http://cython.org/`) has become the principle way for creating compiled extension for Numerical Python. It is as well possible to integrate C and C++ code inside the Python programs, for example, by using CPython.

1.3 Programming or Scripting

Often in literature one reads about *Python scripts*. We will, however, work with programs and not with scripts. What is the difference?

Generally scripts are not structured, but simply a list of instructions that operate on a dataset organized ad hoc for the script. Programs are more autonomous, create own data structures, allocate memory, and in general take more control of the machine. As we will see more in detail in the next chapter using NumPy, Cython, and analogue tools we will take full control of the machine as with a standard programming language, literally *programming* in Python.

The file containing a Python program ends always with *.py* . Python programs are not compiled, but execute instruction by instruction (which ends at the end of line or at a the next semicolon). Python programs, however, are also parsed before execution to check whether evident syntactic errors already exist, and will not run until they are fixed.

1.4 Python Interfaces

In this book I will either show examples from the *Enthought* edition of Python, *Canopy* (`https://www.enthought.com/products/epd/`) or from *Anaconda*, *Spyder* (`https://pypi.python.org/pypi/spyder`), both with a similar graphical style to MATLAB. Each of these interfaces are based on a 'three windows' system in which one has an (i) editor, an (ii) object/variable explorer, and a (iii) standard or iPython console. By using these three at the same time, on a large

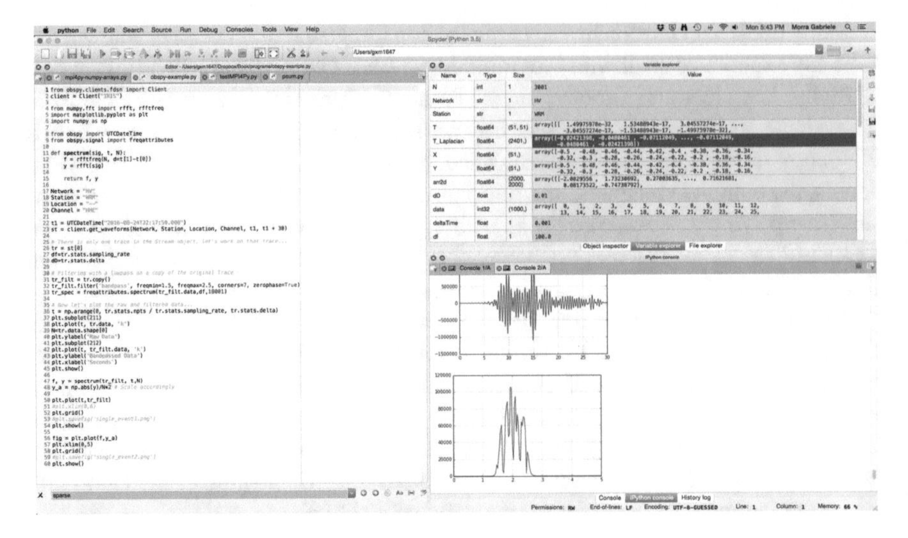

Fig. 1.1 Screenshot of Spyder, the development environment of Anaconda. On the *left* there is the editor. On the *right* on the *top* the variables created in this Python session and on the bottom this *IPython* session with the graphical output

screen, writing, debugging, and testing Python programs is very fast and practical. Fig. 1.1 illustrates a screenshot of Anaconda Spyder.

Installing the *Enthought* and *Anaconda* distributions on a Linux or OSX terminal can be easily done using *bash* or *pip*. Graphical installers exist for both Windows and OSX. Most of the examples shown in this book have been created by using the free version of *Anaconda*.

The easiest way to learn Python and also to begin a new project is to use *iPython*, a project created by the programmer Fernando Perez in 2001 while he was a graduate student in particle physics at the University of Colorado Boulder. During his research Perez realized that the existing tools for scientific computing were insufficient and that Python was the perfect base to start to build new ones. Since then he has been involved in the development of the interactive version of Python, that is today the most used approach to prototyping and experimenting new procedures.

1.4.1 IPython: Interactive Python

When you are ready to learn new computational techniques the question is always what is the fastest and less painful way to start. A straightforward approach is to use *iPython* from the *Anaconda* distribution. After installing the libraries from `https://www.continuum.io/` one can start the iPython with the command:

```
$ ipython
Python 3.5.2 |Anaconda 4.2.0 (x86_64)| (default, Jul  2 2016, 17:52:12)
Type "copyright", "credits" or "license" for more information.

IPython 5.1.0 -- An enhanced Interactive Python.
?         -> Introduction and overview of IPython's features.
%quickref -> Quick reference.
help      -> Python's own help system.
object?   -> Details about 'object', use 'object??' for extra details.
```

Since Python is a relatively new programming language, its tools have been often built in a way to resemble existing well-known software. Aesthetically the *iPython* prompt looks similar to Mathematica. *iPython* has also a number of properties that make it a very practical testing tool, together with its twin environment *Spyder*. People who used MATLAB will find *Spyder* and *Canopy* very familiar.

The online help of *iPython* is very clear and self-explanatory. If this is your first experience with Python I suggest you to play with it and test commands such as:

- Type *help()*, or `help('for')`, or look for other instructions
- `print('I like Geodynamics!')`
- Assign values to string, integer and float variable types.
  - `In[1]: a=1; print(a)`
  - `In[2]: a=1.0; print(a)`
  - `In[3]: a='a'; print(a)`
- Make some mathematical operations (*,/,+,-,**) using the interpreter as a calculator. Notice how python automatically decides which one of the three types of variable, integer, float and string must the answer be.
  - `In[4]: 2+2`
  - `In[5]: 2+2.0`
  - `In[6]: 3**3`
  - `In[7]: 3.0**3`
  - `In[8]: 6/2`
  - `In[9]: 2/6`
  - `In[10]: 'a'+'b'`
- You can also change the type of each variable:
  - `In[11]: int('32')`
  - `In[12]: str(3.2)`
  - `In[13]: int(5.99)`
  - `In[14]: int('5.99') \#this will give you error`
  - `In[15]: int(float('5.99')) \#this works`

If this is your first programming experience, reflect on the answer that *Python* is giving to you. Programming languages have, like in the algebra that we learn at school, the equivalent of *integer numbers*, of *real numbers*, and of *text strings*, therefore its answers reflect these categories, that in Python are called integer (`int()`), float (`float()`), and string (`str()`). Python has always to choose one among them and the interpretation of your instruction will depend to this choice. This is different from languages such as C or Fortran, where you define to start with the type that you are using.

1.5 Few Words on Syntax

An original and practical characteristic of Python (compared to other programming languages) is that it uses the indentation to determine the body of a loop or a function. It is custom to define your functions at the beginning of the file, or in a separate file and import it at the beginning of the main body. You can start with the following example:

```
In[20]: def printSomething():
         print('I like Geodynamics!')
In[21]: printSomething()
```

Here the `print()` instruction is part of the function `printSomething()` while the call `printSomething()` is not indented, therefore it calls the above function.

Functions can be nested one inside another:

```
In[22]: def repeatPrint():
         printSomething()
         printSomething()
```

and inside themselves (recursive functions):

```
In[23]: def fact(n):
         if n==1:
             return 1
         else:
             return n*fact(n-1)

In[24]: print(fact(20))
2432902008176640000
```

Notice how the level of indentation indicates to which loop every instruction belongs. This choice, different from any other language, makes Python closer to human written language, without the large amount of opened and closed parentheses of the C language.

In standard Python one does not need to define the type of every variable. This means that a function will operate differently depending on the type used as input.

```
In[25] : def doublePrint(a):
            print(a+a)

In[26] : doublePrint(2)
Out[26]: 4
In[27]: doublePrint('ciao')
ciaociao
```

The `if` instruction can become a sequence of alternative when combined with `elif`. The standard `or` and `and` logical conjunctions apply:

```
In[28]: def congratulateGrade(grade):
         if grade=='F':
            print('Mmm, something went wrong')
         elif grade=='D':
              print('Not the best performance')
         elif grade=='C':
              print('It is ok!')
         elif grade=='B':
              print('Very good!')
         elif grade=='A':
              print('Fantastic!')
```

Several types of loops exist in Python, however we will exclusively use the `for` instruction. `for` in Python iterates through a list that can contain any kind of object, also belonging to different types. For example,

```
In [29]: for x in [6,'ciao',8*8,congratulateGrade]: print(x)
s6
ciao
64
<function congratulateGrade at 0x10df1d7b8>
```

The last print refers to the function defined above. More in general we will use the `for` loop combined with `range()` or `np.arange()` instructions. For example,

```
In [30]: sum=0
In [31]: for x in range(1000):
            sum+=x
In [32]: print(sum)
Out[33]: 499500
```

As in *C*, `Break` exits a loop and continue to the next iteration. `Pass` does nothing. Another loop instruction that is sometimes used is `while do`. A much more detailed tutorial through all the control flow instructions is available at `http://docs.python.org/3.3/tutorial/controlflow.html`.

To summarize, some key points about programming in Python are:

- There is not an instruction to begin or end the program
- Indentation is meaningful (it indicates whether you are under a condition or in a loop)
- Case matters (i.e., 'PYTHON' is different from 'PyThOn')
- Extra empty lines can be added (the indentation will be the one of the last non-empty line!)
- Comments are inserted starting with the hash symbol '#'

1.6 Extending Python

Besides being dynamic, interpreted, and interactive Python is also Object-Oriented (OO) and extensible. *Object-Oriented Programming (OOP)* is a programming paradigm based on the idea that *objects*, and not *procedures* are at the center of a program, as in traditional programming. *OOP* is based on the idea of *Classes*, from which *Objects* are instantiated. The procedures are then defined only as *methods* associated to each Class and then inherited by objects, or by subclasses, which share some properties with the original Class, and possibly have new methods. The entire concept of programming changes in *OOP* as in it objects interact with each other.

While initially *OPP* was mainly used for static programming, the most important of them being *C++*, in recent years many dynamic programming have been built based on the OOP concept, being *Python* the first of them. *OOP* plays a particular role in Python as it allows its exceptional *extensibility*, or in other words Python can be used to *glue together* a vast range of existing software, developed in a variety of languages, that otherwise could be forgotten, or not used, although they had been written with great care and in order to be very efficient. It was Guido van Rossum in person who pointed this out in 1998 [110].

The use of *OOP* in numerical modeling, in particular geodynamic modeling, until present has been modest. This is mainly due to the fact that geodynamic modeling software requires solving complex and challenging mathematical formulations of geoscientific problems, which do not immediately require being written in *OOP*. However in the case of Python its *Modular* structure allows several important advantages. The first being the possibility to replace some inefficient Numerical Procedure in Standard Python with the better ones of Numerical Python (*NumPy*), but also to import fast libraries initially written in other languages. Let us see how this can be easily done.

1.6.1 Importing Libraries

Every library, or module, can be imported with the command `import`, or a function can be extract from a model by using `from`. If a command exists with the same name, it will be overridden by the new one. For example, let us look at the generation of an array of 10 integers using the module NumPy:

```
In[16]: from numpy import *
In[17]: a=arange(10)
In[18]: print(a)
[0 1 2 3 4 5 6 7 8 9]
```

`arange()` belongs to the *numpy* library, however this is not self-evident, and makes reading the source not immediate. The solution to this ambiguity is to import NumPy as a separate module:

```
In[19]: import numpy as np
In[20]: b=np.arange(12)
In[21]: print(b)
[0 1 2 3 4 5 6 7 8 9 10 11]
```

The advantage of the first method is compactness and speed in writing your script, however the second approach is more clean and helps the reader who might not be familiar with each library to know where each instruction is coming from. Furthermore, the risk of involuntarily redefining a function called with the same name in two libraries is avoided. For this reason in this book we will always use the second form.

1.7 NumPy: Numerical Python

For scientists, Python represented the arrival of a programming language that finally integrates power and ease of use in a nearly optimal way. Its syntax is so simple to be comparable to a scripting language. Its enforced code structure makes it ideal for learning programming for scientists who learn their first language. However, initially the main obstacle to a wide implementation was its lack of performance for large problems.

The *game changing* innovation has been the arrival of numerical libraries that, if properly used, bring up its efficiency to the one languages like C and Fortran. In this book I will often use the words *C-speed*, in analogy to the speed of light, to mean that using specific libraries one can achieve computing efficiency comparable to the language *C*, that like *c* is the unsurpassable speed of physical objects, represents the optimal speed for standard programs.

With the right mix of knowledge between how Python, and in particular its numerical libraries NumPy, uses the low-level computer memory it is possible to make Python performing like compiled codes, avoiding long implementations that are difficult to read and long and tedious debugging. Until now most scientists have achieved a result comparable to that using MATLAB, which is however a proprietary software, therefore with the limits imposed by the developers of the company and requiring a license to run. Python, on the contrary, is an open-source project and is freely available to everyone. While we write Python is scaling up globally and becoming the *de facto* programming language in science.

Specifically in geophysics, and geodynamics in particular, there is a lack of textbooks showing how Python can be used for specific applications. It is my intention to fill this gap with this book, searching for the right balance between introducing geophysicists to programming and teaching them to build geodynamics software.

We will look at the features of *NumPy* in depth in Chap. 3. To show only a first example, we have seen in the Sect. 1.5 that in standard Python one can loop through different objects, and one can call functions with different *variable types*. While this is in general a power feature and has been designed to imitate human language, this is also what makes Python really slow. Every time the parser processes a new variable, it has to *understand* what type it is and then use the right method. Also with very long lists the access of elements is slow due to the unknown size of every variable which makes its access consequential, i.e., O(n) where n is the size the list.

These inefficiencies do not exist in standard compiled languages such as C and Fortran where instead of lists there are arrays of equal types, which allow both processing every variable and accessing every element at the speed of O(1). It also allows extremely efficient storage. *NumPy* offers these properties within a Python framework, leaving the remaining programming experience light and straightforward.

1.8 Visualization

One of the largest projects in Python is the development of the visualization library *MatPlotLib*, available from `http://matplotlib.org/`. While, as it is clear from its name, the initial goal was to create a plotting suite that would imitate the features of MATLAB, with time it gained momentum and is now an independent project and the main visualization library within Python.

While MatPlotLib is a very powerful tool in 2D, in 3D its potential is limited. For this reason for visualizing 3D data and modeling results the best option is to export the data in formats readable by powerful external software such as *Paraview* and *VisIt*. Visualization will be treated in more detail in Chap. 2.

Summary

- History of Python. Python was created in 1989. The version 3.0, released in 2008, was back-incompatible, so new programs should be written with Python 3.x.
- Python is Object Oriented. This allows its highly modular structure. Software can be organized in Classes.
- Python is Extensible. Libraries written in Python or in other languages such as C and Fortran can be efficiently and immediately imported as *Modules*.
- Python is Easy to use. The syntax is extremely simple, compact, and intuitive. The interactive version of Python, *iPython* allows testing codes just on time.
- Modeling with Python. Most numerical outcomes are obtained by using Numerical Python (NumPy), Cython (Python code automatically translated in C) and libraries written in other languages.
- Visualization in Python. Matplotlib is the main visualization library. Python interfaces with VTK, Binary, and other storage and visualization formats.

Problems

1.1 To gain familiarity with the output functions of Python, write a program that writes a sentence on
(a) the screen
(b) on a text file
We did not look yet at how to open and close a file, but it is a simple task that can be found on the online Python documentation, and is therefore a good chance to familiarize with it. Take into account that a file can be opened in *writing* mode, or *reading* mode, and of course you will need the first one. Python 2.x and Python 3.x handle input and output differently, so be aware of which Python version you use be consistent.

1.2 Arithmetic, Geometric, and Harmonic means are very important and different ways to average small or large sets of numbers. We will encounter these types of averaging many times, for example, when averaging the properties of a large number of particles in a cell in the *Particles in Cell* method, or when projecting the solution of a Meshless method to a 3D Lattice in space. Briefly they are defined as:

- Geometric mean: $A = \left(\prod_{i=1}^{n} a_i\right)^{\frac{1}{n}} = \sqrt[n]{a_1 a_2 \cdots a_n}$
- Arithmetic mean: $A = \frac{1}{n}\sum_{i=1}^{n} a_i$
- Harmonic mean: $H = \frac{n}{\frac{1}{x_1}+\frac{1}{x_2}+\cdots+\frac{1}{x_n}} = \frac{n}{\sum_{i=1}^{n}\frac{1}{x_i}}$.

Your task is to write a Python program that generates n random numbers (you can use `a=np.random.rnd(n)`), for example, 10, or 1000 numbers, calculates the three means, and verify which one is systematically the greatest and which one the smallest. Reflect on why it is so. We will use this property.

1.3 Write a program that from an integer number calculates its factorial. There are several approaches for this task. It can be done with a *for* loop, but also with a more elegant recursive function, i.e., a function that calls itself. You can also allow an external user to input the initial number, in which case you can use `Try` and `Except` to handle the possibility that the input is not a positive integer.

1.4 Write a program that calculates the arithmetic mean and standard deviatoric mean of a set of random numbers (google about the deviatoric mean, if you are not familiar with it!). Uniformly distributed random numbers can be generated in Python using the `uniform` instruction within the `numpy.random` module introduced in Sect. 2.2.

Chapter 2
Visualization

Create your own visual style... let it be unique for yourself and yet identifiable for others.

—Orson Welles

Abstract Examples are introduced here in which Python can be used to visualize Scientific data in 2D and 3D. Data can originate from both observation and numerical models. Examples in this chapter focus more on observational data. The most used visualization library in Python, *MatPlotLib*, is introduced from the historical point of view. Datasets are downloaded from the web, the length of the day through several years. A mathematical 3D object called *Moebius stream* is created and visualized with several techniques. Python is mostly used as a wrapper for general visualization software such as *Paraview* and *VisIt*. More in general there exists a *Scientific Python Ecosystem*, embedding tools such as *iPython* (now *Jupyter*), MatPlotLib, and many more. This is a set of tools that allows practicing, visualizing, and sharing technologies applied to Science.

When Python was initially created, the main software used in the academia and many private data analysis/numerical modeling companies was MATLAB. For this reason, it became natural to the developers who worked on a visualization library that would work with Python to create a library with an environment that was as similar as possible to MATLAB. In this way, scientists would have found natural to transition to the open software.

As in many other Numerical Python projects, the initiator was a scientist. In this case the Neuroscientist John D. Hunter, who did not find an open-source option for visualizing his data, the Electrocorticography of patients with epilepsy, contacted Fernando Perez, the creator of iPython, in order to add a specific patch to iPython. As Perez himself recalled, however, Perez was too busy writing his Ph.D. dissertation, which brought John D. Hunter to just create a new library from scratch [107]. This library was called *MatplotLib*, and remained his main project until his sudden death in 2012. Since then a team of developers continues to expand the library.

Although *MatplotLib* can plot some features in 3D, most of its power lies on its 2D capabilities. Practically every datasets that can be visualized in 2D can be efficiently

G. Morra, *Pythonic Geodynamics*, Lecture Notes in Earth System Sciences,
DOI 10.1007/978-3-319-55682-6_2

and beautifully visualized with Matplotlib. For example, almost all the figures of the now historical paper that announced the discovery of the Gravitational Waves have been done with Python and MatplotLib [111], and more in general Numerical Python is used daily by the LIGO and VIRGO projects, in control room, Signal Processing, parameter estimation, etc.

In 3D, instead, only few Python-based softwares exist, and Python can be employed more proficiently as a wrapper to collect data, organize them, and finally interface with the major existing visualization software. I will cover here two of the main open source software available: Paraview and VisIt.

Visualization is not the most difficult problem to solve in Geodynamics, certainly not as much as the parallel programming of multi-scale nonlinear Stokes of Darcy flow equation, however without an efficient and effective visualization framework it is impossible to quickly understand the outcome of the development efforts themselves. Ultimately, it is essential for every modeler to have a clear understanding of what the options available in visualization and how to quickly implement them.

From this point of view, visualization also enters into the more general problem of scientific reproducibility. Two projects, heavily based on MatPlotLib, that address this are *iPython* and its newer version, *Jupyter*. Here Visualization is intended in its most general form, as representation of a scientific result. To use iPython Notebooks or Jupyter Notebooks to reproduce a theoretical or numerical work allows (i) giving all the details on its reproducibility and (ii) checking its validity. Many years ago a scientific result was validated by its reproducibility by the calculation on some pieces of paper. With the advent of computing many numerical and data-based results cannot be manually reproduced anymore. Following many before me, I strongly advocate for the adoption of a general framework of openness in data and procedures behind every scientific result, and I believe that the Python powered scientific environment will play a key role in the next future of this revolution.

2.1 The MatPlotLib Visualization Library

Psychologists have proven that more than 50% of our attention is captured by our visual input. This implies that the development of the superb, flexible, and stable MatPlotLib library has played an essential role in the development of the Scientific Python project. *MatplotLib* is today the most commonly used visualization library in Python. It was started by John D Hunter [72], who has built most of it by himself, until he passed away in 2013, and left an enourmous imprint to the Python community of developers.

There are fundamentally two ways to call the *MatplotLib* functions: (1) the *PyLab* interface that facilitates the transition of experienced MATLAB users into *MatplotLib*, whose development was the initial goal of John Hunter, and (2) a python interface that can be used in a much more general framework. The quickest and smartest way to make a new 2D or 3D plot is to look at the vast gallery of graphs available at `http://matplotlib.org/gallery.html`, each with the associated Python scripts. In most cases one can find a figure close to the style in which one wants to show data or results and use it as a template. Let us look now at some easy examples.

2.1.1 Plotting a 2D Field

The best way to learn how *MatplotLib* works is by playing with the examples available in the online *gallery*. Let us, for example, assume that we know a function z of x and y: $z = f(x, y)$. In this case, we assume that $f(x, y)$ is the sum of two Gaussian functions, called in 2D *Bivariate Normal*. We will center them one in $(-2, -2)$ and the other in $(1, 1)$, with sizes $(1, 2)$ and $(2, 1)$. These functions are already implemented in Python:

```
import numpy as np
import matplotlib.pyplot as plt

x = y = np.arange(-5.0, 5.0, 0.1)
X, Y = np.meshgrid(x, y)

Z1 = plt.mlab.bivariate_normal(X, Y, 1.0, 2.0, -2.0, -2.0)
Z2 = plt.mlab.bivariate_normal(X, Y, 2.0, 1.0, 1.0, 1.0)
Z = (Z1 - Z2)
```

Here after importing *NumPy* and *MatPlotLib*, I define a background regular mesh. On this *lattice*, I calculate the Gaussian Normal function and operate on them. To plot the difference Z between the Gaussians with a filling contour plot, I need the intuitive command:

```
CS = plt.contourf(X, Y, Z)
```

where however the however requires a *colorbar* to be better interpreted:

```
CS = plt.contourf(X, Y, Z); plt.colorbar(CS)
```

the result is on the left of Fig. 2.1. A large number of options exist to plot more accurately a 2D field with *MatPlotLib*. For example to add contour lines, define values and write them on the plot itself one can program:

```
levels = np.arange(-0.095,0.095,0.02)
CS = plt.contourf(X, Y, Z, levels)
CS4 = plt.contour(X, Y, Z, levels, colors='k')
plt.clabel(CS4, colors='k', fontsize=16)
```

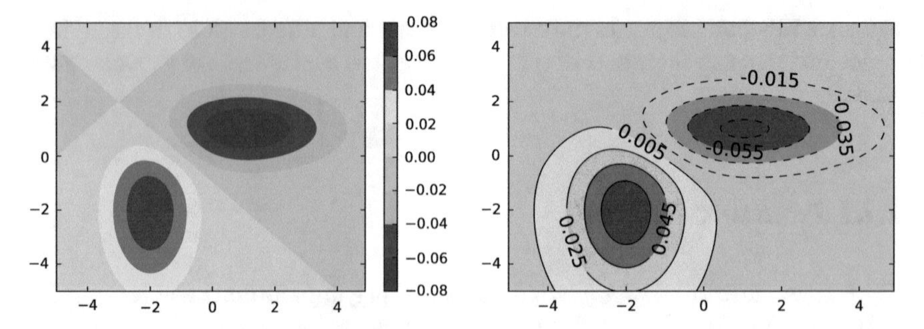

Fig. 2.1 *Left* filled contour plot of the difference of two bivariate normal distributions with a colorbar on the side. *Right* the same functions plotted now with labeled values on the contour boundaries

which gives the result on the right of Fig. 2.1. A figure like this can be saved and shown on the screen with:

```
plt.savefig('filename.pdf')
plt.show()
```

The sequence is important. If `plt.show()` is shown before, `plt.savefig()` will save a blank image.

2.1.2 Plotting a Map

In Geodynamics we often want to map our numerical models on the Earth surface, or plot our data on a physical Map. Aimed at doing this MatplotLib has a specific module called *mpl_toolkits.basemap*. This can be installed, e.g., in Anaconda with the command *conda install basemap*. Let us first create a very simple projection of a map on a sphere of a simple function (decays exponentially toward the poles, oscillated like a wave with the longitude). First we load the module, create the coastlines, color the continents versus the oceans and plot the latitude and longitude:

```
import mpl_toolkits.basemap as bm
import matplotlib.pyplot as plt
import numpy as np

# Let us set our view above Eurasia. And use low resolution. myMap =
bm.Basemap(projection='ortho',lat_0=30,lon_0=60,resolution='l')

# draw coastlines, countries and continents.
myMap.drawcoastlines(linewidth=0.25)
myMap.drawcountries(linewidth=0.25)
myMap.fillcontinents(color='orange',lake_color='aqua')
```

```
# draw the edge of the map, the meridians and the parallels
myMap.drawmapboundary(fill_color='aqua')
myMap.drawmeridians(np.arange(0,360,30))
myMap.drawparallels(np.arange(-90,90,30))
```

You can already run this script in *iPython* and visualize it with `plt.show()`. We are now ready to add a function on top, and I will create them by adding a baseline to a wave on a 100×100 grid (I use here some *NumPy* functions that I will explain in detail later):

```
# make up some data on a regular lat/lon grid.
n = 100; d = 2*np.pi/(n-1)
[lats,lons] = d*np.indices((n,n))
wave = np.exp(-lats/10) * np.cos(lons)
mean = np.cos(2.*lats) * np.sin(2.*lats)
x, y = myMap(lons*180./np.pi, lats*180./np.pi) #project the lat/lon on the grid
cs = myMap.contour(x,y,mean+wave,15,linewidths=1.5) #project a contour plot on
  ↪ the map
plt.title('Example of a plot over a global map')
plt.show()
```

And one obtains the result shown in Fig. 2.2.

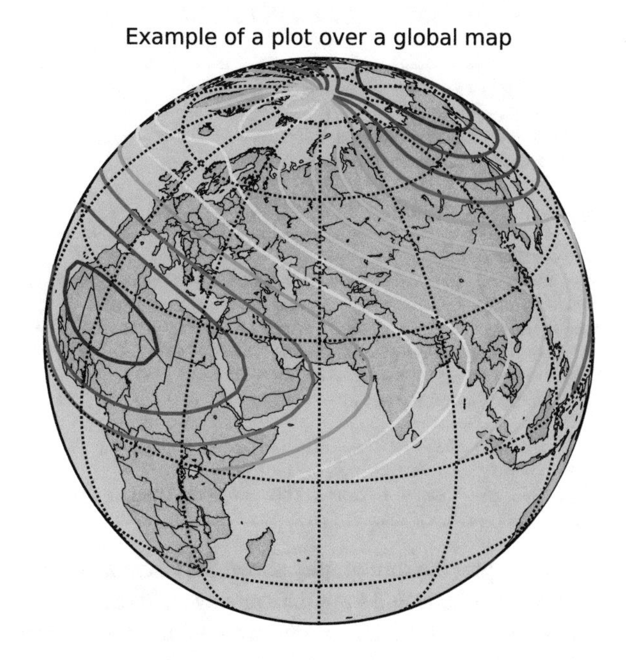

Fig. 2.2 Example of a simple made-up function plot over a spherical projection of the Earth

2.1.3 NetCDF and ETOPO

In the next example, I will use the standard storage system called *NetCDF4*. NetCDF stands for Network Common Data Form and it is an interface for storing and retrieving data in form of arrays, that are the main format of data that we use in our models and for data representation. To store the data in arrays has numerous advantages that we will exploit when learning how to use NumPy. Among them that array values may be accessed directly, ignoring the details of how the data are stored. This means that netCDF datasets can be only partially accessed and transformed, without locally storing and managing the entire datasets, which therefore can be very large.

The physical representation of netCDF data is designed to be independent of the computer on which the data were written. In particular in the version four of NetCDF the modern data format HDF5 was introduced, allowing unlimited datatypes and specifically designed for efficiently transmit high volume and complex data. Examples on how to create, store and retrieve HDF5 data in Python are in `https://support.hdfgroup.org/HDF5/examples/intro.html#python`.

In order to learn to use NetCDF4, dataset let us play with them and plot a public topography/bathymetry dataset of *NOAA*. First, we have to access it. For this exercise you can also download the entire dataset, but it is huge, while the idea of accessing NetCDF data is to extract only the data that we are interested in. We will use the database of the software *Ferret*. In Anaconda, the netCDF library can be installed with the command *conda install netCDF4*.

```
import mpl_toolkits.basemap as bm
import numpy as np
import matplotlib.pyplot as plt
from netCDF4 import Dataset

#loading the data and extracting the latitude, longitude and topography
#if this dataset is not available anymore, it can be downloaded from many
  ↪ sources
#as well as the more recent higher resolution versions
url = 'http://ferret.pmel.noaa.gov/ thredds/dodsC/data/PMEL/etopo5.nc'
topography = Dataset(url) #extract data using NetCDF
topoin = topography.variables['ROSE'][:]
lons = topography.variables['ETOPO05_X'][:]
lats = topography.variables['ETOPO05_Y'][:]

# ETOPO and basemap are shifted of 180 degrees in longitude, so we need to
  ↪ shift the reference
topoin,lons = bm.shiftgrid(180.,topoin,lons,start=False)
```

Let us plot the topography above the Himalaya, without country boundaries, but only with latitude and longitude plot every 20 degrees. The key function that creates the map is again *basemap*, that we now use with parameters to define the plotting window (min and max lat and lon) and viewpoint from the space.

```
fig = plt.figure()
ax = fig.add_axes([0.1,0.1,0.8,0.8])
m = bm.Basemap(llcrnrlon=40.,llcrnrlat=0., urcrnrlon=140,urcrnrlat=60.,\
          resolution='l',area_thresh=1000., projection='lcc',\
          lat_1=30.,lon_0=90.,ax=ax) #viewpoint from space

n = 1000
nx = 1 + int( (m.xmax-m.xmin)/n )
ny = 1 + int( (m.ymax-m.ymin)/n )
topodat = m.transform_scalar(topoin,lons,lats,nx,ny)
# plot image over map with imshow.
im = m.imshow(topodat)
m.drawcoastlines()
par = np.arange(0.,80.,20.); m.drawparallels(par,labels=[1,0,0,0])
mer = np.arange(10.,360.,20.); m.drawmeridians(mer,labels=[0,0,0,1])
cb = m.colorbar(im,"right", size="5
ax.set_title('Topography above Himalaya')
plt.show()
```

And one obtains the result shown in Fig. 2.3.

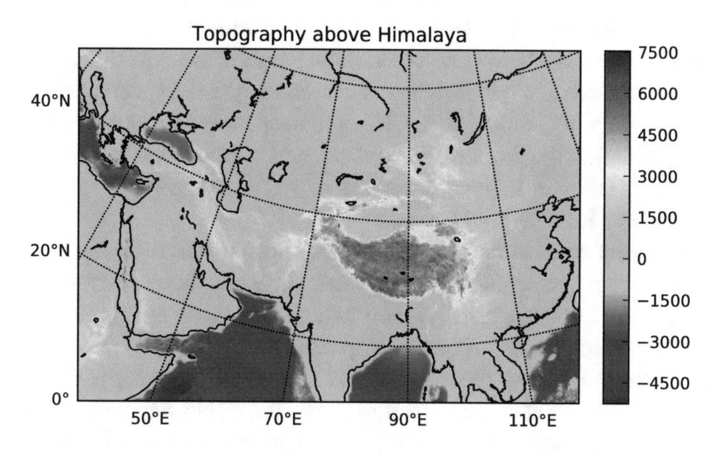

Fig. 2.3 Plot of the topography map over the Himalayan region using NOAA public Data

These were really two very rudimentary examples. Many more and more sophisticated ones can be found in the webpage of the *basemap toolkit* of *MatplotLib*.

There are many other tools for plotting Maps. For example, *PyFerret* allows to extract and manipulate the above dataset with greater speed and care. And the most used Plotting tool in Geophysics, *GMT* can be used as well, through the binding of *PyGMT* and *GmtPy*, although they are not so updated. While I write there is a new project, called gmt-python, lead by Leonardo Uieda and his supervisor Paul Wessel, who promises to become a definitive binding package. *GMT* commands can be also called from the shell line within Python. For example, all the plots in the publication [113] have been made in this way.

There are really advanced projects for global visualization of data related to plate tectonics reconstruction, with the most prominent one being probably *GPlates*, that has been written in many parts using Python and allows interfacing with external scripts by using Python, as explained in its tutorial at `https://www.gplates.org/docs.html`. Although GPlates presently only plots maps on the two-dimensional surface of the Earth, plans exist for its extension to visualizing it in combination with the Earth's interiors.

2.1.4 Plotting a Seismic Waveform

One of the most inspiring stories of geophysics refers to Bernard Chouet. Born in Switzerland, after graduating in Electrical Engineering, was hired at MIT to work on the NASA's Apollo's missions, therefore at the age of 30 took a Master in Geophysics and started looking at the seismic signatures of volcanoes, with the ambition of predicting the behavior of volcanoes. After obtaining a Ph.D. again at MIT he moved to USGS, where he worked until he discovered in the seismic records of the eruption of 1985 in Nevado del Ruiz, Colombia, that by filtering the seismic signals of the volcano for long-period waves, he could detect and increase of the activity before the eruption. He used then this insight to successfully predict several other eruptions in Alaska and again in Colombia. The life of many people has been saved in Mexico in 2000 by predicting the eruption of Popocatepetl in Mexico using his method.

Today thanks to open Internet databases of worldwide seismic records, we can repeat his analysis in many locations.

The community of seismology made a great effort to deliver extensive data based that could be used by seismologists to test their software, their algorithms, and build seismological and tomographic models. I will show here an example from the *IRIS* database. *IRIS* stands for *Incorporated Research Institutions for Seismology* and its data can be accessed using the *client* of its Python library *obspy*. The following code plots a seismic event at the Hawaii in 2016. The events happened near Kilauea, Hawaii and was detected at the *West Rim* site. I show here a simple filtering of the seismogram where only the frequencies between 1.5Hz and 2.5Hz are selected and used. the results is shown in Fig. 2.4. Again, there is not enough space to list here all the plotting possibilities of *obspy*, furthermore these libraries evolve very rapidly, so the best way to learn to use them is to access the tutorial that for *obspy* is located at the address `https://docs.obspy.org/tutorial/`.

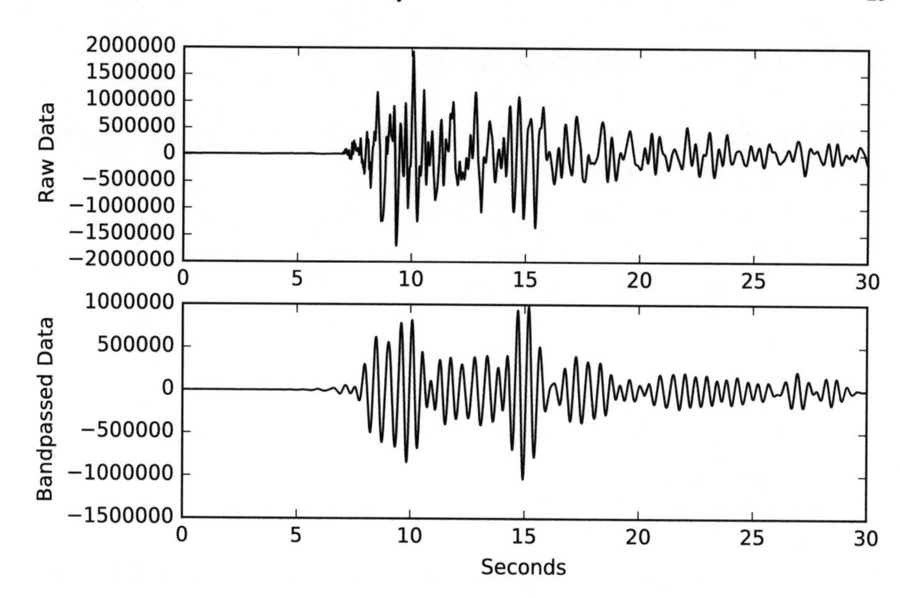

Fig. 2.4 Seismogram of an event on August 24, 2016, detected at the West Rim, Hawaii. The above plot shows the original data, and the one below the filtered one

```
#Plotting a seismic event near Kilauea
import matplotlib.pyplot as plt
import numpy as np
from obspy.clients.fdsn import Client
from obspy import UTCDateTime
from obspy.signal import freqattributes

# Load the data from the web
Network = "HV"; Station = "WRM"; Location = "--"; Channel = "HHE"
t1 = UTCDateTime("2016-08-24T22:17:50.000")
client = Client("IRIS")
st = client.get_waveforms(Network, Station, Location, Channel, t1, t1 + 30)

# There is only one trace in the Stream object, let's work on that trace...
tr = st[0]; df=tr.stats.sampling_rate; dD=tr.stats.delta

# Filtering a copy of the original Trace
fMin=1.5; fMax=2.5
tr_filt = tr.copy()
tr_filt.filter('bandpass', freqmin=fMin, freqmax=fMax, corners=3,
  ↪ zerophase=True)
tr_spec = freqattributes.spectrum(tr_filt.data,df,18001)

# Plottnig raw and filtered data
t = np.arange(0, tr.stats.npts / tr.stats.sampling_rate, tr.stats.delta)
plt.subplot(211); plt.plot(t, tr.data, 'k'); plt.ylabel('Raw Data')
plt.subplot(212); plt.plot(t, tr_filt.data, 'k')
plt.ylabel('Bandpassed Data'); plt.xlabel('Seconds')

plt.show()
```

2.2 Plotting in 3D with MatPlotLib

MatplotLib allows also plotting in 3D. For example, a set of 1000 randomly placed points can be displayed with the script:

```
import numpy as np
import matplotlib.pyplot as plt
from mpl_toolkits.mplot3d import Axes3D
import matplotlib.cm as cm

n = 1000
fig = plt.figure()
ax = fig.add_subplot(111, projection='3d')
color=cm.rainbow(np.linspace(0,1,n))
xs = np.random.rand(n)
ys = np.random.rand(n)
zs = np.random.rand(n)
ax.scatter(xs, ys, zs, c=color)
plt.show()
```

One outcome is in Fig. 2.5. Colors are assigned sequentially using the function `plt.cm.rainbow()`. From this simple snipped, one can see how the 3D plots can be rotated and seen from any angle.

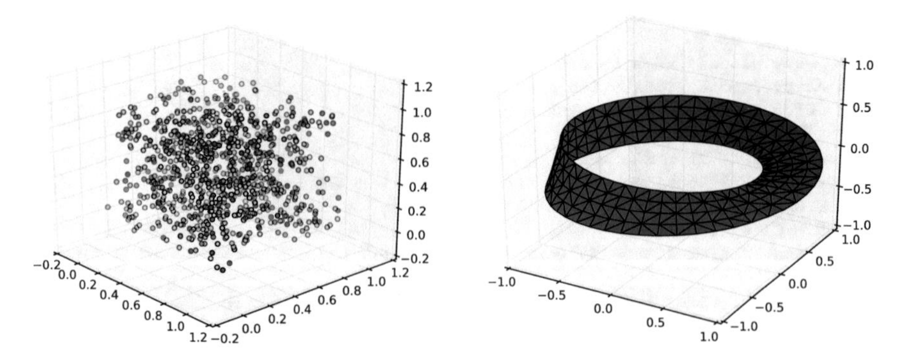

Fig. 2.5 *Left* 1000 dots randomly placed in the space. *Right* the Moebius strip obtained using the `tri.Triangulation()` function of MatPlotLib

2D surfaces in the 3D space can be plot as well. For example, the celebrated Möbius strip, which is a surface with only one side and only one boundary, can be created by the 2D to 3D mapping:

$$
\begin{aligned}
x(u, v) &= \left(1 + \frac{v}{2}\cos\frac{u}{2}\right)\cos u \\
y(u, v) &= \left(1 + \frac{v}{2}\cos\frac{u}{2}\right)\sin u \\
z(u, v) &= \frac{v}{2}\sin\frac{u}{2}
\end{aligned}
$$

I show here how to redo it with Numerical Python. Some of the tools (`np.arange(); np.ones(); np.reshape()`) that I employ in the snipped below will be explained in Chap. 3. Let us instead focus on the Matplot lib functions.

One is *matplotlib.tri.Triangulation*, that takes a collection of points in the 2D space and turns it into a triangulated mesh. If the mesh is not explicitly indicated, Python will build the Delaunay mesh, which is a triangulated mesh that maximizes the size of the angles of the mesh, i.e., avoids thin and long triangles. This is an extremely useful tool that we will use again.

```
import numpy as np
import matplotlib.pyplot as plt
from matplotlib import tri

n = 50; l = 5

u = np.arange(n)*2*np.pi/(n-1)
u = u * np.ones((l,1))
u = u.reshape(n*l)

v = np.arange(l)/(l-1)-0.5
v = v * np.ones((n,1))
v = v.transpose().reshape(n*l)

# Mobius mapping, (u, v) -> (x, y, z)
x = (1 + 0.5 * v * np.cos(u / 2.0)) * np.cos(u)
y = (1 + 0.5 * v * np.cos(u / 2.0)) * np.sin(u)
z = 0.5 * v * np.sin(u / 2.0)

# Create the triangulated surface
surface = tri.Triangulation(u, v)

fig = plt.figure()
ax = fig.add_subplot(111, projection='3d')
ax.plot_trisurf(x, y, z, triangles=surface.triangles)
ax.set_xlim(-1, 1); ax.set_ylim(-1, 1); ax.set_zlim(-1, 1);
```

In the above example, n and l are the parameters controlling the density of the mesh. You can play with their values if their meaning is not immediately clear.

MatPlotLib plots triangulated surfaces by using the `matplotlib.plot\_trisurf()` function. A view of the Möbius stream created by the above program is in Fig. 2.5.

2.2.1 VTK File Format

While in 2D Matplotlib allows plotting almost anything, its 3D features are limited, in particular when dealing with large amounts of data, as is often the case in 3 dimensions. For this reason very large models can be better visualized using third part software. The most common tool used in geosciences is probably *Paraview*, and open and free software, which also runs in parallel when the size of the dataset requires it. For simple datasets as the one above, it is possible to simply use ASCII or binary *VTK* files.

VTK stands for Visualization Toolkit and is associated to a large number of popular data file formats. In *VTK* there are mainly two different styles of file formats. The classical one is the simplest and consists in a serial list of points, connectivity, data. This structure has the advantage of being easy to read and write, normally with a program, and even by hand. We will look here only into this format. An alternative is the *XML* format, that supports random access and parallel input/output. For more information on this last format, the *VTK* website `www.vtk.org` is very well documented.

Here below, I show a routine that allows creating a *VTK* file with any surface created with `matplotlib.tri.Triangulation()`:

```
def writeVTKSurface(outFileName,x,y,z,triangles):
    nodesNumber=x.size
    trianglesNumber=int(triangles.size/3)
    m=open(outFileName,'w')
    m.write('# vtk DataFile Version 2.0\n')
    m.write('Moebius surface\n')
    m.write('ASCII\n')
    m.write('DATASET POLYDATA\n')
    m.write('POINTS '+str(nodesNumber)+' float\n')
    for node in np.arange(nodesNumber):
        m.write(str(x[node])+' '+str(y[node])+' '+str(z[node])+' \n')

    m.write('POLYGONS '+str(trianglesNumber)+' '+str(trianglesNumber*4)+'\n')
    for triangle in np.arange(trianglesNumber):
        m.write('3 '+str(triangles[triangle,0])+'
  ↪  '+str(triangles[triangle,1])+' '+str(triangles[triangle,2])+' \n')

    m.close()
    return()
```

This function can be called in our case, for example, with the command:

```
writeVTKSurface('moebius.vtk',x,y,z,surface.triangles)
```

2.3 Example: Length of the Day

The ability to find and process datasets is essential in geophysics. I show here an example on how to download a dataset from the web (evolution of the Length of the Day for one year) from an official repository and how to plot these data using MatPlotLib.

Earth–Moon dynamics, variations of the Earth's inertial axis, seasonal oscillations, local tides, and other phenomena make the Earth's rotation rate slightly vary in time, causing a fluctuation of the Length of the Day. Present GPS data allow estimating the Earth's rotation rate every day, and therefore the Length of the data (LOD) on a daily bases.

As an example, I illustrate here a plot of a recent dataset from the International Earth Rotation and Reference System Service (`www.iers.org`). I download first the raw data from January 2016 that are available at the address `http://datacenter.iers.org/eop/-/somos/5Rgv/latestXL/207/bulletinb-337/csv`. *csv* is a format for data storage and in our case it is in *ascii*, therefore we can simply parse the file to find necessary data.

In Python 3.x the important Module *urllib* has been introduced, which allows requesting, downloading and writing a file online, if permitted. We will use this module to download the data. Generally in *csv* files the fields are divided by *,* or *;* or a similar symbol. In our case, a direct inspection of the data file shows that it is the symbol *;*. Furthermore the first line of the *csv* file shows the meaning of every field and that the first field is the day and the Length of the Day is the thirteenth, which are indexes 0 and 12 in Python.

We will store the cardinal day number and the Length of each Day in the lists *day* and *LOD*. It is important, when reading an unknown file, to allow the possibility that one line could be not read. This is normally done with the instructions `try` and `except` that control this case and avoid an error that would irremediably stop the execution:

```
import urllib.parse
import urllib.request
import matplotlib.pyplot as plt

day=[]; LOD=[]
url='http://datacenter.iers.org/eop/-/
  ↪ somos/5Rgv/latestXL/207/bulletinb-337/csv'

with urllib.request.urlopen(url) as response:
lines = response.read()
lines=str(lines)
lines = lines.split('\\n')

for line in lines:
    words = line.split(';')
    try:
       (thisDay, thisLOD)=( float(words[0]),  float(words[12]) )
         day.append(thisDay); LOD.append(thisLOD)
    except:
        print('line not readable')
```

```
plt.plot(day,LOD)
plt.show()
```

Here we have extract the data from the file as lines of text and then used the function *split()* to divide different fields. Different lines are divided by the *newline* special character (*n* in our case). Different fields by *;*. Finally, we transform the text into float numbers and append to the initially empty *day* and *LOD* lists. The plot associated to this script is on the left of Fig. 2.6.

To visually detect cycles in the evolution of the LOD we need to look at several months, or better years. This can be done by extending the above script to reading a series of files, one for each month, and then putting all the data together. For example the following could be a way (only the modified part is shown):

```
numberOfYears=5
day=[]; LOD=[]
fileNames=[]
url = 'http://datacenter.iers.org/eop/-/somos/5Rgv/latestXL/207/'
for i in range(338-12*numberOfYears, 338, 1):
    fileNames.append('bulletinb-'+str(i))

for fileName in fileNames:
    with urllib.request.urlopen(url+fileName+'/csv') as response:
        lines = response.read()
        lines=str(lines)
        lines = lines.split('\\n')
        for line in lines:
            words = line.split(';')
            try:
                (thisDay, thisLOD)=( float(words[0]) ,  float(words[12]) )
                day.append(thisDay); LOD.append(thisLOD)
            except:
                print('line not readable')
```

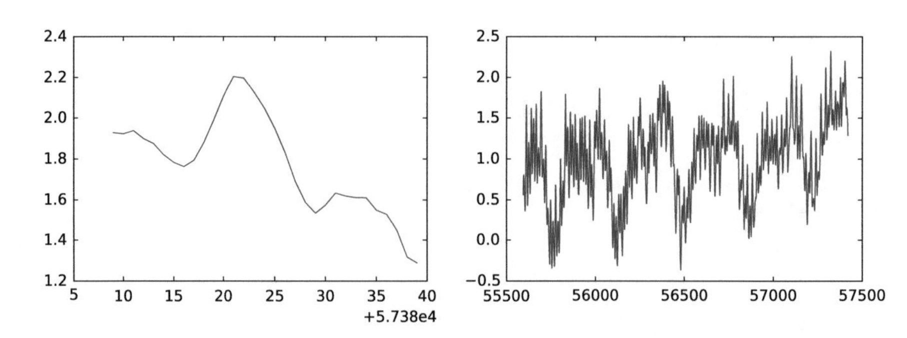

Fig. 2.6 *Left* time series extracted from the *IERS* website related to the variations of the Length of the Day in January 2016. *Right* the same plot for a period of 5 years, combining the data extracted by 60 (5*12) monthly files from the same source. In this second plot at least two overlapping cycles are clearly visible

The result of this script is on the left of Fig. 2.6 and now several cycles are clearly visible. We will see in the next chapter, using NumPy how we can extract more information from this time series.

2.4 IPython and Jupyter Notebooks

IPython, *Interactive Python*, initially started as a very practical and handy interface for Python, in particular aimed at prototyping new programs. Today it is a growing, language-agnostic, project. Agnostic in the sense that it is based on a *Notebook* format. To learn iPython is like to learn MatPlotLib: the best way is to start from examples, which can be all found presently at the page `ipython.org`. An even greater set of examples can be found in the *A gallery of interesting IPython Notebooks*, presently available on the `github.com` platform [108].

Among the many features of iPython, it is important to mention the existance of the *magic commands* that allow to quickly perform *extra Python* operations. These commands are of two types, either *line magics* or *cell magics*, the first being anticipated by % and the second by %%. Example of the first case is *%time*, that allows testing the speed of a certain command. For example, to benchmark the time necessary to create an array of sequentially 1 billion integers one can write *%time sum(range(1000000000))* (27 s on my laptop). Examples of the seconds are *%%script*, that allows writing a shell command and possibly run other languages (e.g., *%%script ruby*) inside a Python script.

In 2015, the iPython project has evolved into the Jupyter project, where the *Notebook* is now at the center. This application allows creating and sharing documents in the form of a *Notebook*. These documents can contain live code, equations, visualizations, and explanatory text. The idea behind their creation is extremely ambitious and in many ways aims at transferring the framework of the Open-Source Software (OSS) into Academia. Jupyter Notebooks have the characteristic of being very technical and detailed. In this sense, they allow to distribute knowledge in a new form compared to technical publications where only general directives about implementations are given, allowing to increase the speed at which science can spread.

Jupyter Notebooks do not exist only for Python but with many other languages as well, however here I will show only examples with Python. On printing this book, I plan to transfer most of the examples from this book into Jupyter Notebooks and to share them online.

The reason for inserting *iPython* and *Jupyter* in the visualization chapter is that these tools aim at reproducing and sharing techniques. They are therefore a way to *visualize* them.

2.5 Paraview and VisIt

One of the best Python integrated powerful visualization software, openly and freely available to anyone, is ParaView. Personally it has been my preferred choice for over a decade. Paraview is multiplatform, working on Linux, OS-X and Windows. On every platform it builds 3D visualizations at an amazing speed plotting any kind of data on volumes, surfaces, sections and other more complex integrated geometries. Normally it is used in an interactive mode, to visualize ongoing simulations, however using Python is possible to programmatically process it and create figures with a consistent outlook, which is particularly useful when writing publications, e.g. to use the same color-scale and plot 3D data from the same angle for large datasets.

The name ParaView clearly comes from Parallel, because it has been developed in order to analyze extremely large datasets on a distributed computing system. It can run on supercomputers and analyze petabytes of data, well beyond what can be stored in the hard-drive of most desktops, but it results extremely fast and efficient on a laptop as well. A detailed and updated guide to Paraview is available at `http://www.paraview.org/paraview-guide/` in PDF. Here I only show few examples to illustrate its potential.

The simplest way to combine Python and Paraview is to create VTK files and open them with Paraview. For example the Moebius stream that I created above can be visualized immediately (see Fig. 2.7).

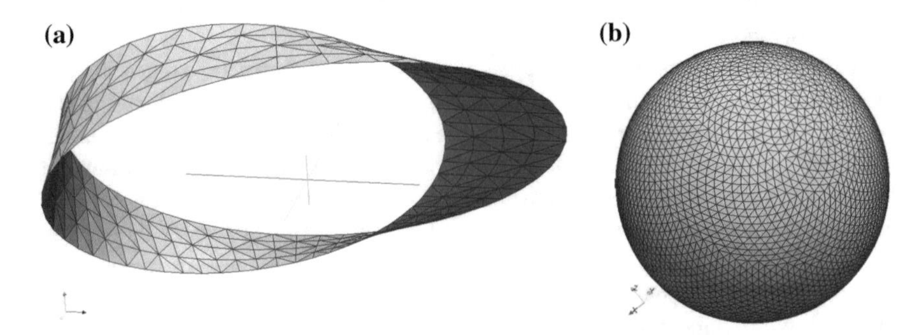

Fig. 2.7 Left: Snapshot of the visualization of the Moebius stream done with Paraview. Right: visualization of the mesh of a sphere obtained with a Fibonacci spiraling technique, again visualized from a VTK file in Paraview

One can similarly create much more complex setup and visualize results. For example I can create a set of bubbles in space, at random position and of random size (with certain limits), that has been used for research (e.g., [27, 123, 124]. Over there, the mesh was created by recursively subdividing the faces of a regular polyhedron in 4 parts until the desired resolution is achieved. I show here instead a simpler approach based on spiraling around a sphere to setup the points of the mesh and then using the *Convex Hull* to triangulate the surface:

```
import numpy as np
from scipy.spatial import ConvexHull

samples = 5000
increment = np.pi * ( 3 - 5**0.5 )
points = np.zeros((samples,3))

phi = np.arange(0., samples*increment, increment)
points[:,1] = ( 2. * np.arange(samples) + 1 )/samples - 1.
r = ( 1 - points[:,1]**2)**0.5
points[:,0] = np.cos(phi) * r
points[:,2] = np.sin(phi) * r

hull = ConvexHull(points)
```

Here the key is in using the interval equal to the angle in radiants $\pi(3 - \sqrt{5})$, derived from the golden number, and obtaining a sequence like the Fibonacci one. This creates a perfect covering of a sphere, with any given number of number. The result of this mesh is shown in Fig. 2.7. In Exercise 2.5 it is suggested to use a number of spheres created with this technique to create volume filled with spherical beads. Techniques to create such filling, important in many fields of science of heterogeneous media, and in many fields of petrology and sedimentology, can be found in [45].

Alternatively one can create a Python script and run it with *pvpython*, the dedicated python interpreter created for Paraview. In *pvpython* both NumPy and MatplotLib can be called, making it a powerful tool, however most of its command are complex and its strict object oriented structure make it difficult to use for the beginner programmer.

Another large open source project in which Python plays a big role is *VisIt*. VisIt, like ParaView, is designed for post processing of mesh based data, mostly scientific. Python was adopted early by the VisIt team as the VisIt's primary scripting interface [73]. VisIt's developers have embedded Python interpreters into their data flow network pipelines, allowing users to write custom algorithms in Python to manipulate the data in the mesh. Once installed *VisIt*, any python program can be sent to *VisIt* just with the command *./visit -nowin -cli -s <script.py >*. On the *VisIt* platform many examples are given on how to plot functions and dataset from VTK sets. Generally the outcome is similar to Paraview, with however more attention to visualizing fluid-dynamics, streamlines and movies.

2.6 Python as a wrapper: SEATREE and Underworld

Large collaborative projects specifically applied to geodynamics also exist. Many computational geodynamicists use commonly Python as the wrapper for integrating existing open source software. *SEATREE* (Solid Earth Teaching and Research Environment) is general project that aims at connecting in the most efficient way very different computational modules ranging from a mantle convection code, to flow visualization, from mantle tomography sampling and subsampling to 3D visualization, from body wave mantle seismic tomography to earthquake location inversions. The goal of SEATREE is its application in the classroom, and as a platform for scientific collaboration. Here Python is used mainly as a wrapper to combine together existing well-known codes that have been used by the scientific community for many years.

The purpose of SEATREE is different from most of the examples shown in this book, which have mainly the purpose of understanding the techniques behind geodynamic numerical modeling. SEATREE is instead an attempt to provide an easy and graphically supported *black box* tool, where however what lies under the hood is well explained.

Underworld is also project aiming at combining modeling and visualization, but by using original and innovative Particles in Cell implementations. It is Python-friendly by providing a programmable and flexible front end that allow to quickly setup standard geodynamic setups for running parallel HPC simulations. It is specifically suited for modellers who do not have an in-depth knowledge of the code development and want to focus on complex geodynamic simulations. Tutorials to learn how to use Underworld also use the Jupyter Notebooks, which makes easy to integrate it with the techniques introduced in this book.

Summary

- MatPlotLib is the most used visualization library in Python. It is particularly powerful in 2D, but limited in 3D, particularly for visualizing the results of complex numerical models.
- Maps and other geometrical features are already implemented in MatPlotLib. It is straightforward to plot fields on geographical projections in 2D and 3D.
- MatPlotLib allows plotting features in 0D (points), 1D (segments), 2D (triangles, squares) as well as 3D (spheres, blocks, etc.).
- Paraview and VisIt are much more powerful visualization tools but need specific file formats (such as VTK) to be used efficiently to plot numerical results.

Problems

2.1 Plot the topography of Africa, using the same `http://ferret.pmel.noaa.gov/thredds/dodsC/data/PMEL/etopo5.nc'` of Sect. 2.1.3. The topography data are called *ROSE* and latitude and longitude are called *ETOPO05_X* and *ETOPO05_Y* respectively. By setting to zero all negative topography values and plotting only the `np.log10()` of the topography the continental morphology appears very neat.

2.2 Using the strategy shown in Sect. 2.1.3 for accessing large NetCDF files, extract the global precipitation data from the NOAA website (e.g. `http://www.esrl.noaa.gov/psd/thredds/dodsC/Datasets/cmap/std/precip.mon.mean.nc`) and plot a precipitation map for a continent for Jan. 2009. A careful look at the NetCDF file reveals that the data are organized in *time*, *longitude* and *latitude*, and averaged per month, starting in January 1979, therefore to obtain the data for January 2009 one needs to extract the month after exactly 30 years.

2.3 Extract and visualize the sea surface Temperature from the Sect. 2.1.3 file at the NOAA website (e.g. `ftp://ftp.cdc.noaa.gov/Datasets/COBE2/sst.mon.mean.nc`). The data file is about 500 Megabytes large, therefore extract only the necessary data. The dataset ranges from 1850 to 2015. Make one plot per year to show the time progression.

2.4 Download a NetCDF file of Paleo Age Crustal Data from the `http://www.earthbyte.org/Resources/agegrid2008.html` website. The data are organized in *age*, *longitude* and *latitude* as z, x, y. Plot the seafloor age for an ocean, such as *Atlantic*, *Pacific* or *Indian*. Use the `mpl\_toolkits.basemap` as shown in Sect. 2.1.3.

2.5 Build a VTK file (Sect. 2.2.1) and show in Paraview (Sect. 2.5) the highly compact set of spheres that occupies the volume $1 \times 1 \times 1$ with a compaction of 70%. You will need to use sphere with different dimensions as such a high compaction cannot possibly be achieved with sphere of the same size. Use first large sphere and then fill the remaining space with smaller spheres, where possible. The setup can be all identified before building the mesh, since the distance between two spheres is given by the distance between the centers mines the sum of the two radii. Use Paraview to visualize the sphere surface mesh.

Chapter 3
Fast Python: NumPy and Cython

"Give me 6 hours to chop a tree, I will spend the first 4 sharpening my axe." — *Abe Lincoln*

Abstract Python can combine simplicity with speed in a unique way, but much care has to be paid in avoiding bottlenecks, particularly avoiding unnecessary loops and heavily using its vectorized functions. In this chapter I make a detailed overview of which types exist in Python, how to speedup calculations using arrays, how to efficiently select and operates on subsets of large arrays. By illustrating *strides*, I explain how Numerical Python can achieve its performance and how to benchmark them. Finally, I illustrate how to add vector and linear algebra calculations to an applied modeling software and eventually use Cython, the C extension to Python, to add special functions that cannot be performed using the vectorized NumPy library.

Ultimately computational Geodynamics presently consists in expressing evolutionary laws in terms of differential equations, which we solve in a discretized form. This form consists in most cases in a *lattice*, a cartesian structure that we either use to solve our equations (Finite Difference) or to project the solution obtained with other numerical techniques (Particles method, Surface-based methods). Ultimately we desire to describe these equations as arrays and matrices, and to apply known linear algebra tools to solve them.

We will introduce the necessary equations very gradually. The aim of this chapter is to introduce some concepts of Linear Algebra that we necessarily need to know before starting modeling and to show how they can be naturally described and solved using the Numerical Python (*NumPy*) library. For particularly hard problems, I will introduce at the end of the chapter alternative techniques based on precompiling Python programs as they were written in C (*Cython*), run parallel routines (*mpi4py*), and other strategies.

G. Morra, *Pythonic Geodynamics*, Lecture Notes in Earth System Sciences,
DOI 10.1007/978-3-319-55682-6_3

3.1 How Fast is Your Computing Machine?

Our computers do basically two things, one is storing numbers, the other is doing mathematical operations on them. They are so useful because they do these things extraordinarily fast and extraordinarily well, in the sense that they always do it correctly following our instructions. For this reason the main way in which we use them is to give them repetitive instructions for up to billions of times.

Since the speed of the CPU has duplicated ever 1.5 years until few years ago (Moore's Law), we may have lost track of how really fast is the computer in our hands. When starting this journey into numerical modeling, it is important to first quickly experiment this and realize how remarkably fast it really is, but also to perceive its limits when we reach it.

In this chapter, we start using iPython because its interactive setting allows us *feeling* better what the computer does. Later we will step into writing and running a program. Let us initially load NumPy:

```
In [1]: import numpy as np
```

and let us test how large has to be an input number *max* in order to start noticing how long it takes to repeat and extremely basic task, like counting.

```
In [2]: def count(max):
...:      i=0
...:      while i<max:
...:          i=i+1
...:      print(i)
...: I
```

When I try to launch *count(1000)* or *count(100000)* on my laptop I simply do not observe any difference. Just the time to call the function, and I get the result. It is unnoticeable that the second call required 100 times more calculation than the first one. Only when I arrive at one million or better ten millions, I finally observe a lag. In fact what I notice is the time that it takes to one processor of my laptop for this serial calculation. You can test yours and *perceive* the power of your computer.

We can be also more quantitative about the time required by our little routine. *iPython* offers a very simple and powerful *magic command* called `%timeit`. Let us test it for testing the time required by our loop:

```
In [3]: %timeit count(100000)
100 loops, best of 3: 11.9 ms per loop

In [4]: %timeit count(1000000)
10 loops, best of 3: 114 ms per loop

In [5]: %timeit count(10000000)
1 loops, best of 3: 1.09 s per loop
```

The way in which `%timeit` operates is by repeating the calculation several times and then choosing the best. The reason for this strategy is to avoid that side processes might load the processor and overestimate the calculation time. The number of repetitions will be greater if the time required for each loop is very small (e.g., 100 times for 10 ms).

It is inspiring to play a bit with loops and `%timeit` before proceeding toward more advanced topics in this book. For example, it is interesting to look at what happens to the performance of the `count` function by replacing `while i<max` with `for a in range(max)`. The performance improves or not? Why?

Another instructive test is to calculate `a=a+i` instead of `i=i+1`:

```
In [6]: def addCount(max):
   ...:     a=0
   ...:     for i in range(max):
   ...:         a=a+i
   ...:     print(a)
   ...:
```

Comparing the two versions (with `a=a+i` vs. `i=i+1`) shows that the performance is the same. This depends on the fact that the sum operation applied to two integers takes the same time, regardless of the quantities that are added. By replacing `a=0` with `a=0.`, therefore implicitly assuming that a is a real and not an integer, that instead requires 20–30% longer to run. These simple exercises helps gaining a better *feeling* about what makes our software performing better and what does not.

3.2 Numerical Python

The main tool used for high-performance computing with Python is NumPy, short for Numerical Python. Nearly all the examples in this book are based on this library (or Module) that, when properly used, allows gaining one to two orders of magnitude in speed compared to standard Python. This chapter is mostly about how NumPy works and how to increase the speed of our calculations.

Among others, NumPy adds to standard Python the following features:

A. Multidimensional vectorized arrays
B. Mathematical functions operating on an array or portions of it
C. Linear Algebra, Fourier development, Random Functions
D. Input/Output functions to efficiently create and read memory mapped files

In the rest of this chapter, we will always assume that the NumPy package has been loaded (`import numpy as np`), so that every NumPy function *xxx* will be called as *np.xxx*. Alternatively NumPy functions, as for every other Module, can be loaded overwriting other Python functions with `from numpy import*`.

3.2.1 NumPy Types

It is said that the flattering of a butterfly in Brazil can cause a hurricane in the US. Similarly poor handling of a rounding error in programming can produce a terrible disaster in the real life. This is for example what happened on February 25, 1991, when a Patriot missile failure in Dharan, Saudi Arabia, was responsible of 28 deaths. Similarly, the overflow due to the erroneous conversion from a 64-bit floating point value to a 16-bit signed integer value was ultimately the cause of the explosion of an Arianne 5 rocket just after lift-off in French Guinea, on June 4, 1996 [13].

Without any further specification, Python automatically uses float type *machine precision*. We might not know which precision this is, in which case it is important to discover it. A very straightforward way to discover the precision of our machine is by testing for which small ε, a and $a + \varepsilon$ become indistinguishable. We can do so by iteratively decreasing ε of a certain $ratio$:

```
epsilon, ratio = 1.0, 9.0
while (epsilon):
    print('Precision:',epsilon)
    epsilon /= ratio
```

On my laptop, after many iterations, this gives me finally the output:

```
[...]
Precision: 3.9503855941862293e-305
Precision: 4.389317326873588e-306
Precision: 4.877019252081764e-307
Precision: 5.418910280090849e-308
Precision: 6.021011422323166e-309
Precision: 6.6900126914702e-310
[...]
Precision: 2.132e-320
Precision: 2.367e-321
Precision: 2.6e-322
Precision: 3e-323
Precision: 5e-324
```

The last line indicates that the precision that standard Python assumes on my machine is 10^{-324}. This is indeed associated to a 64-bit float number. One can verify this by setting a smaller precision in the first line of this sequence. Setting initially:

```
epsilon, ratio =  np.float16(1), np.float16(9)
```

Iterations stop at $1.e-7$:

```
Precision: 1.0
Precision: 0.11108
Precision: 0.012344
Precision: 0.0013714
Precision: 0.00015235
Precision: 1.6928e-05
Precision: 1.9073e-06
Precision: 2.3842e-07
```

Similarly to float, there exist integers with several precisions: *np.int32*, *np.int64*, *np.int128*. The type of every NumPy variable can be retrieved the *dtype*, e.g., *eps.dtype*.

It is very important to keep in mind the importance of precision. All the mathematical routines implemented in Python, as well as in any other language, follow the standards determined by IEEE which imply that they are all corrected to the last digit. Therefore, when we will model a physical system we can be sure that errors will not propagate quickly to lower digits with every algorithm that does not explicitly remove this information.

3.2.2 ndarrays

Let us start looking at the vectorized arrays. The core of NumPy is the object called *ndarray*, a storage of large quantities of data that allows to operate fast and flexible operations on it. Let's look at some basic examples using *iPython* again and create a matrix (a two-dimensional array) of two rows and three columns:

```
In [7]: firstArr = np.array([[4,6,3],[2,3,6]])

In [8]: firstArr # shows the matrix
Out[8]:
array([[4, 6, 3],
       [2, 3, 6]])

In [9]: firstArr.ndim #  number of dimensions
Out[9]: 2

In [10]: firstArr.shape # array shape as (rows,columns)
Out[10]: (2, 3)

In [10]: firstArr.dtype # Out[10]: dtype('int64')
```

What we find is that NumPy, if given a certain set of data, automatically decides a type (generally integer or float) and defines an array of the right shape where to store them. Given a NumPy array, it is always possible to obtain its shape and data type (*dtype*). The main difference between *ndarrays* and Python data types is that the elements of *ndarrays* are all of a predefined and homogeneous type, which is one of the reasons for the speed of its calculations.

In general, one can set the data type at the moment of creation of the array, which is a particularly useful function when one has to handle very large models and datasets and needs to be in control of the size of the occupied memory. Data in NumPy are either *int*: integer, *uint*: unsigned integer (from 0), and *float*, a real number. *int* and *uint* can have sizes of 8, 16, 32, and 64 bits, while *float* of 16, 32, and 64 bits. If not otherwise set, automatically NumPy will set data type size to 64 bits.

Most commonly used generators of *ndarrays* are `arange`, `zeros` and `ones`. Let us look at some examples:

```
In [11]: secondArr = np.arange(20)

In [12]: secondArr
Out[12]:
array([ 0,  1,  2,  3,  4,  5,  6,  7,  8,  9, 10, 11, 12, 13, 14, 15, 16,
17, 18, 19])

In [13]: secondArr.dtype
Out[13]: dtype('int64')

In [14]: secondArr = np.zeros(20)

In [15]: secondArr
Out[15]:
array([ 0., 0., 0., 0., 0., 0., 0., 0.,0., 0., 0., 0., 0., 0., 0., 0.,
0., 0., 0., 0.])

In [16]: secondArr.dtype
Out[16]: dtype('float64')

In [17]: secondArr = np.ones(20)

In [18]: secondArr
Out[18]:
array([ 1., 1., 1., 1., 1., 1., 1., 1.,1., 1., 1., 1., 1.,  1., 1., 1.,
 1., 1., 1., 1.])

In [19]: secondArr.dtype
Out[19]: dtype('float64')
```

We see that `arange` is the analogue of `range` in Python, but it creates a NumPy array of 64 bits integers. `zeros` and `ones` instead create arrays of 64 bits float. `ones` can also be created by `zeros` using the broadcasting properties of NumPy arrays, just by the instruction `arr=np.zeros(20)+1.0`. If we wish to create arrays with different types, for example due to operational reasons or of memory size, dtypes can be explicitly specified, e.g., `arr = np.arange(20, dtype='float32')` or `arr = np.ones(20, dtype='int8')`.

Let us now check how using NumPy allows to speed the calculations of the prior section. The first temptation might be to simply to replace *range* with *np.arange*.

```
In [20]: def addCountNP(max):
...:     a=0
...:     for i in np.arange(max):
...:         a=a+i
...:     print(a)
...:
```

Doing so, however, one observes that the time necessary to run the program is even longer than the one required by standard Python. Let us run our tests on arrays of the size of 100 millions numbers:

```
In [21]: %timeit addCountNP(100000000)
1 loops, best of 3: 16.6 s per loop

In [22]: %timeit addCount(100000000)
1 loops, best of 3: 12.1 s per loop
```

What happened here is that although the arrays were *ndarrays*, the operation was done as for a standard Python list. The trick to speed calculations on *ndarrays* is to use the broadcast vectorized version of each operation. Let us look at how to add two *ndarrays* of one million of integers either with the standard Python loop and exploiting the NumPy broadcasting capabilities:

```
In [24]: def addArray(a,b):
....:     c=np.zeros(a.size)
....:     for i in np.arange(a.size):
....:         c[i]=a[i]+b[i]
....:     return(c)
....:

In [25]: a=np.arange(1000000)

In [26]: b=np.arange(1000000)

In [27]: %timeit c=addArray(a,b) #standard python
1 loops, best of 3: 639 ms per loop

In [28]: %timeit c=a+b #NumPy arrays broadcasting
100 loops, best of 3: 3.74 ms per loop
```

A gain of two orders of magnitude! To use the broadcasting feature of NumPy makes Python's speed comparable to compiled codes such as C, but with the obvious huge gain in terms for code development, testing, and readability. Broadcasting is clearly not limited to the addition operation, but it works as well with all the other arithmetic operations such as `c=a*b; c=a/b; c=a-b`, as well as scalar-array operations like `c=1/a; c=a**0.5`.

Besides operations on two arrays, there are numerous unary (and binary universal functions that can be used with NumPy arrays. Among unary functions, much used ones are *np.abs(); np.sqrt(); np.exp(); np.log(); np.sign()* and all the trigonometric functions. Binary functions are *np.add(); np.multiply(); np.power(); np.maximum(); np.mod()*.

Let us go back now to the initial problem. We wanted to sum all the elements of a large array (100 millions numbers). This is neither an unary or binary operation, because it is a function that projects an array into one number, the sum. The most common among these operations are already efficiently implemented in NumPy. For example, the sum is immediately obtained by:

```
In [23]: %timeit np.arange(100000000).sum()
1 loops, best of 3: 531 ms per loop
```

These operation have been in fact written in Cython, which is a C compiled operation. We will see later in this chapter how such operations, when not already implemented in NumPy, can be easily created using Cython.

3.3 Indexing and Slicing

The reader that has some experience in programming knows that most languages today are *compiled*. One of the great features of compilers is that when we have to perform a set of repetitive operations on a large set of data, like the temperature of on a cartesian network of points in the space that we use to describe a solid, the compiler will take care to optimize these iterations in an exceptional way, even we wrote a very inelegant code. This is not the case with an interpreted software like Python.

If we want our Python code to run fast, we have to organize the sequence of operation in a smart way. In particular a very common case like calling an *if* command inside a *for* loop can make Python very slow, as we have seen how looping makes Python slow. So the secret that makes Python fast is to use smart indexing. To use indexing in practice in this case means to vectorize the command *if* and then to use these indexes to selectively operate on arrays. This might seem a terrible setback for Python programmers, however the great advantage of forcing the developer to do this operation is that the program will be written in a completely vectorized way from the start, therefore ready to be parallelized.

Indexing and slicing in NumPy is a long topic, whose full coverage goes beyond the scopes of this book. To gain a full understanding of the possibilities offered by NumPy I recommend to follow one of the online free tutorials (e.g., `https://docs.scipy.org/doc/numpy-dev/user/basics.indexing.html`). I will cover here some main the features that we will use more often in this book, and explain in depth some important details on the memory management associated to *ndarrays*.

Let us now dive into the main features that interest us. Given an array, e.g., *arr*, we can access the element *n* with square brackets *arr[n]* and we can *slice* the array extracting the elements between n and m with the command *arr[n:m]*, which will return $m - n$ elements (comprising *arr[n]* and excluding *arr[m]*. It is very important to understand that even if we associate a name to the slice of *arr*, this is only a *view*, not a new array implying that the $m - n$ elements are not copied in a new allocated chunk of memory. This means that by changing the sliced array you will change the initial data. Let us look at an example to clearly understand how this works:

```
In [29]: secondArr=np.arange(20)

In [30]: secondArr
Out[30]:
array([ 0, 1, 2, 3, 4, 5, 6, 7, 8, 9, 10, 11, 12, 13, 14, 15, 16, 17, 18, 19])

In [31]: sliceArr = secondArr[4:10]

In [32]: sliceArr
Out[32]: array([ 4,  5,  6,  7,  8,  9])

In [33]: sliceArr[3]=100000

In [34]: sliceArr
Out[34]: array([ 4,  5,  6,  100000,  8,  9])

In [35]: secondArr
Out[35]:
array([ 0, 1, 2, 3, 4, 5, 6, 100000, 8, 9, 10, 11, 12, 13, 14, 15, 16, 17, 18,
  ↪ 19])
```

Who is familiar with other languages might be surprised by this behavior, and believe that when defining *sliceArr* Python should have copied the subset of *secondArr* data into a new array. The point is that NumPy has been designed to deal with very large datasets or numerical models, therefore it uses a policy of minimization of memory usage. For this reason, if one does not force NumPy to create a copy of the sliced data, it will simply generate a *view* of the already existing array. NumPy can be forced to create a separate new array by adding *.copy()* to the slice, e.g., in our case *sliceArr = secondArr[4:10].copy()*.

Slicing is very flexible and allows omitting, for example, the first index (e.g., *arr[:10]*) or the last index (e.g., *arr[10:]*) of a slice, which implies that the slice reaches the end of the array. It is also possible to slice the array every k elements by indicating a third parameter in the slice (e.g., *arr[4:12:3]*), and also omitting the other parameters (e.g., *arr[::3]*). Negative indexes are also admitted, which means that counting starts from the last element, backward, for example:

```
In [36]: secondArr[10:-5]
array([10, 11, 12, 13, 14])
```

3.3.1 N-Dimensional Indexing

Indexing and slicing in more dimensions are just a recursive repetition of one-dimensional operations. In 2D, for example, arrays can be indexed either as *arr2d[n][m]* or, with the same effect, as *arr2d[n,m]*. The first index refers to the inner array, the second index to the outer array. For example,

```
In [37]: arr2d = np.arange(20).reshape((5,4))

In [38]: arr2d
Out[38]:
array([[ 0,  1,  2,  3],
       [ 4,  5,  6,  7],
       [ 8,  9, 10, 11],
       [12, 13, 14, 15],
       [16, 17, 18, 19]])

In [39]: arr2d[2]
Out[39]: array([ 8,  9, 10, 11])

In [40]: arr2d[2,3]
Out[40]: 11

In [41]: secondArr[4*2+3]
Out[41]: 11
```

It is very important to remember that the *reshape* command does not create a copy of the data, as well as slicing. The data remain stored in the memory as a one-dimensional array, regardless of the dimensions and shape of the array. The way in which this is done and how NumPy broadcasting operations remain extremely efficient is explained in the Sect. 3.4. Also a book that analyzes in greater detail how Python can deal with data is [25].

3.3.2 Boolean Indexing

Boolean indexing is the fastest and efficient way to select, access and operate on subsets of a NumPy arrays. Let us for example create a random 5×5 matrix (array) and select only the positive elements:

```
In [42]: arr2d = np.random.randn(5,5)

In [43]: arr2d
Out[43]:
array([[ 1.61431976,  0.15569749, -2.14364165,  0.38966765,  0.85815266],
[-0.36096076, -1.64827638,  1.75104872,  0.07810726,  0.24062888],
[ 0.28342606,  1.17805318,  0.02322372,  0.63254705, -0.74592907],
[-0.66752419, -0.62794643,  0.03543305,  0.25110785, -1.22947746],
[-0.34985565, -2.2051168 , -0.3946053 , -0.81539967, -0.57438627]])
```

```
In [44]: arr2d>0
Out[44]:
array([[ True,  True, False,  True,  True],
[False, False,  True,  True,  True],
[ True,  True,  True,  True, False],
[False, False,  True,  True, False],
[False, False, False, False, False]], dtype=bool)

In [45]: arr2d[arr2d>0]
Out[45]:
array([ 1.61431976,  0.15569749,  0.38966765,  0.85815266,  1.75104872,
0.07810726,  0.24062888,  0.28342606,  1.17805318,  0.02322372,
0.63254705,  0.03543305,  0.25110785])
```

What we did was to create a 5×5 array whose values were *True* when the elements were positive, and *False* in the opposite case. It was then possible to select only those elements. It is important to notice that the extracted elements have lost their 5×5 structure. This again results from the fact that Python stores every array, regardless of its shape, as a 1D array.

This technique is normally used to operate on a certain subset of an array. For example if we desire to set to 0 all the negative elements of the above array, we can do it with one instruction:

```
In [46]: arr2d[arr2d<0]=0

In [47]: arr2d
Out[47]:
array([[ 1.61431976,  0.15569749,  0.        ,  0.38966765,  0.85815266],
[ 0.        ,  0.        ,  1.75104872,  0.07810726,  0.24062888],
[ 0.28342606,  1.17805318,  0.02322372,  0.63254705,  0.        ],
[ 0.        ,  0.        ,  0.03543305,  0.25110785,  0.        ],
[ 0.        ,  0.        ,  0.        ,  0.        ,  0.        ]])
```

It is important to emphasize that besides compactness and elegance, this technique guarantees an enormous gain in speed. To quantify this advantage, we can compare the time necessary for a standard Python *for loop* with the Boolean indexing tool described above for a large 2000×2000 random array:

```
In [48]: arr2d = np.random.randn(2000,2000)

In [49]: def setNegativeValuesToZero(n,m,a):
    .....: for i in np.arange(n):
    .....:  for j in np.arange(m):
    .....:      if a[i,j]<0:
    .....:          a[i,j]=0

In [50]: %timeit -n1 -r1 setNegativeValuesToZero(2000,2000,arr2d)
1 loops, best of 1: 2.18 s per loop

In [51]: arr2d = np.random.randn(1000,1000)
```

```
In [52]: %timeit -n1 -r1 arr2d[arr2d<0]=0
1 loops, best of 1: 10.2 ms per loop
```

Again, we observe a gain of 2 orders of magnitude by using the NumPy indexing compared to standard Python loops. The message is now loud and clear: never use Python loops for large datasets and large numerical models, but always employ Python indexing/slicing features to achieve compiled code performance for speed and memory management. When not possible, rely on *Cython* Sect. 3.7.

3.3.3 Transposing and Axis Rotation

We already encountered the function *reshape*, which allows to *project* a 1D array as a generic n-dimensional array. One operation that cannot be achieved by reshaping is however transposing, which corresponds to swapping axis. This is a very important function when operating on 2D and 3D numerical modeling, but it is also simply essential in order to calculate the inner product. For example, to calculate $X^T X$, one writes:

```
In [53]: arr2d = np.random.randn(9,3)

In [54]: np.dot(arr2d.T,arr2d)
Out[54]:
array([[ 22.05210899,  -3.84117516,  -8.96398348],
[ -3.84117516,  10.55000505,   1.09331319],
[ -8.96398348,   1.09331319,  22.83143636]])
```

Transposing, however, is not limited to swapping x and y axis. One can rotate axis as well. For example in 3D:

```
In [55]: arr3d = np.arange(2*3*4).reshape(2,3,4)

In [56]: arr3d
Out[56]:
array([[[ 0,  1,  2,  3], [ 4,  5,  6,  7], [ 8,  9, 10, 11]],
    [[12, 13, 14, 15], [16, 17, 18, 19], [20, 21, 22, 23]]])

In [57]: arr3d.transpose((2,0,1))
Out[57]:
array([[[ 0,  4,  8], [12, 16, 20]],
    [[ 1,  5,  9], [13, 17, 21]],
    [[ 2,  6, 10], [14, 18, 22]],
    [[ 3,  7, 11], [15, 19, 23]]])
```

Transposing, as well as reshaping, is just a specific view of the entire array, therefore transposing will not create a new set of data, and when modifying the transposed of an array, one modifies the original array as well.

3.4 Strides

This section is not essential in order to understand the rest of the book, so if some aspects are not clear, it is possible to just skip it. However, the technical details given here help to understand how NumPy deals with very large arrays minimizing memory occupation and maximizing access speed.

It is important to emphasize that although an array can have many dimensions, the memory in our computer is structured in a purely sequential way, therefore ultimately the real structure of the array is one-dimensional, and every n-dimensional array will only be a specific *view* on that 1D array.

NumPy is extraordinary powerful in managing n-dimensional arrays, different views, complex subsets (e.g., `arr[arr*arr<1]`. This power is based on the ability to directly address chunks of data by *striding* across the memory and *zooming* into small blocks of memory. For example, let us take a random $100 \times 100 \times 100$ array of *np.float64* data, equivalent to a 1D array of one million elements, occupying 8 bytes each. A block within the array can be extracted in microseconds, almost regardless to the size of the block:

```
In [58]: arr3d = np.random.randn(100,100,100)

In [59]: %timeit -n1 -r1 arr3d[45:55,45:55,45:55] #small block
1 loops, best of 1: 7.41 micro-s per loop

In [60]: %timeit -n1 -r1 arr3d[15:85,15:85,15:85] #large block
1 loops, best of 1: 7.61 micro-s per loop
```

When we operate on the subset of the array however the operational time increases with the size of the block, although not in a proportional way. For example, to set the values of the subset to zero we need about 20 times more time for a subset that is over 300 times greater:

```
In [61]: %timeit -n1 -r1 arr3d[45:55,45:55,45:55]=0 #small block
1 loops, best of 1: 28.9 micro-s per loop

In [62]: %timeit -n1 -r1 arr3d[15:85,15:85,15:85]=0 #large block
1 loops, best of 1: 622 micro-s per loop
```

NumPy accesses n-dimensional arrays and its slices as fast as one-dimensional arrays. To understand how this is achieved, let us consider, for example an array *a* of 32000 integers and then create a reshaped three-dimensional $20 \times 40 \times 40$ version assigned to *b*. In our case *a* is composed by 64-bit (8 bytes) integers, which means that one needs to proceed 8 bytes forward to access the next element along the first axis. The size of every element can be also in general assessed with *a.itemsize*:

```
In [63]: a=np.arange(32000)

In [64]: b=a.reshape(20,40,40)

In [65]: print(a.itemsize,b.itemsize)
Out[65]: 8 8
```

This will be stored in memory in a buffer that contains 32000 ascending integers from 0 to 31999. As we have seen before *a* and *b* are stored in the same memory block. The way in which NumPy differentiates how to operate on them is by characterizing them by their different strides. Strides are tuples of bytes to step in each dimension when traversing an array. In practice they are the offset in *bytes* between an element and the neighboring one in every direction. Strides are shown explicitly using the instruction *arr.strides*:

```
In [62]: b.strides
Out[32]: (12800, 320, 8)

In [62]: a.strides
Out[33]: (8,)
```

The last number, 8, of the strides refers to the size of each element in the array, i.e., it is always the *itemsize*. The other numbers refer to the number of bytes forward necessary to access the next element along the other axes. The total number of bytes is called *offset* and is calculated as `offset = sum(indexes * a.strides)`. For example, for the initial array *a[n]* every element is accessed at the position `offset = n*a.strides[0]`, while *b[i,j,k]* is accessed calculating its associated memory offset as `offset = i*b.strides[0]+j*b.strides[1]+k*b.strides[2]` in bytes. To obtain the element location one has to divide by *b.itemsize*. For example,

```
In [62]: b[2,3,4]
Out[35]: 3324

In [62]: (2*b.strides[0]+3*b.strides[1]+4*b.strides[2])/b.itemsize
Out[38]: 3324.0
```

3.5 Vector Products

The inner product, *np.dot()* in NumPy, is the most common operation between arrays. It is equivalent to the inner product between 1-D arrays or to matrix multiplication between 2D arrays. It is defined on n-dimensional arrays as the sum product over the last axis of the first array and the second-to-last of the second array. For example,

```
In [56]: np.dot(np.random.rand(1000),b=np.random.rand(1000))/1000
```

will give you a number close to 0.25, since *np.random.rand()* is a uniformly extracted random number between 0 and 1. Similarly

```
np.dot(np.random.rand(100,100),b=np.random.rand(100,100))/100
```

will result in a 100×100 matrix of numbers normally distributed around 0.25.

Outer products are called by *np.outer()* and are an easy way to create an x, y regular mesh:

```
In [56]: a=np.arange(-10,11,1)
In [56]: a
Out[125]:
array([-10, -9,-8, -7, -6, -5, -4, -3, -2, -1, 0, 1, 2, 3, 4, 5, 6, 7,
 8, 9, 10])

In [56]: b=np.ones(21)
In [56]: b
Out[127]:
array([ 1., 1., 1., 1.,1., 1., 1.,1., 1., 1., 1., 1., 1., 1., 1., 1., 1., 1.,
1., 1.,1.])

In [56]: x=np.outer(a,b)
In [56]: y=np.outer(b,a)
In [56]: plt.plot(x,y,'o')
```

That is showing a 21×21 regular mesh in figure Fig. 3.1.

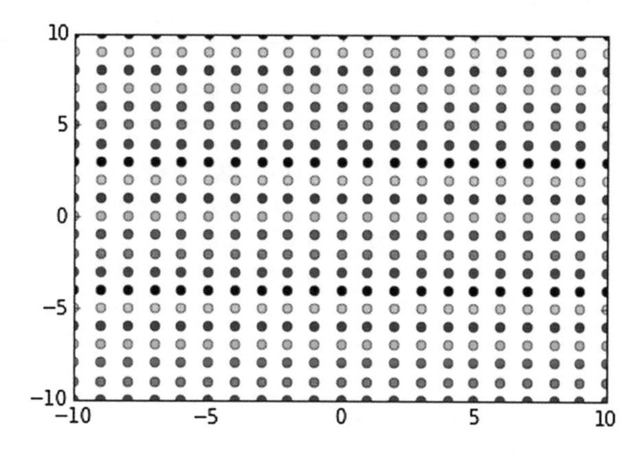

Fig. 3.1 *Left dots* indicating the corners of a regular $x - y$ mesh

NumPy allows composing large arrays combining blocks of one array scaled by another. This operation is called Kronecker product, and we will intensively use it to build large operators for two-dimensional continuum mechanics. For example,

```
In [56]:np.kron(np.eye(3), np.ones((2,2)))
Out[56]:
array([[ 1.,  1.,  0.,  0.,  0.,  0.],
       [ 1.,  1.,  0.,  0.,  0.,  0.],
       [ 0.,  0.,  1.,  1.,  0.,  0.],
       [ 0.,  0.,  1.,  1.,  0.,  0.],
       [ 0.,  0.,  0.,  0.,  1.,  1.],
       [ 0.,  0.,  0.,  0.,  1.,  1.]])
```

3.6 Linear Algebra

Stand firm in your refusal to remain conscious during algebra. In real life, I assure you, there is no such thing as algebra.

NumPy actually supports this famous sentence of Frances Ann Lebowitz, as all the linear algebra tools that we will need are already efficiently implemented through the standard optimized ATLAS LAPACK and BLAS libraries. Ultimately all linear algebra routines expect to operate on a 2-dimensional array. There is also a *matrix* type in NumPy but its use is discouraged since it is possible to obtain the same result by using *arrays* only.

Linear algebra has many subprograms that run routines for decompositions such as *Cholesky*, *QR* and *Singular Value*. They can be very important for many problems, however they do not apply for the problems that we will address in this volume. The interested reader can refer to the regularly updated SciPy manual at `http://docs.scipy.org/doc/numpy/reference/routines.linalg.html`. Useful discussions and more details on the importance of Linear Algebra routines in modeling are also available at [59, 60].

Linear algebra (LA) routines are accessed through the module *numpy.linalg*:

```
In [80]: import numpy as np
In [81]: import numpy.linalg as linalg
```

An often employed tool is *linalg.norm()* both for calculating the size of a 1D (vector) and 2D (matrix) arrays. Norm, that is just the square root of the sum of the square of all the elements of a n-dimensional array, could be also calculated by the definition, but the *numpy.linalg* implementation can be one order of magnitude as more efficient. Let us for example benchmark the calculation of a 100×100 array norm:

```
In [81]: a=np.arange(-100,101,1)
In [82]: b=np.arange(201)
In [83]: x=np.outer(a,b)

In [84]: %timeit linalg.norm(x)
10000 loops, best of 3: 42 micro-s per loop

In [85]: %timeit np.sqrt(sum(sum(x**2)))
1000 loops, best of 3: 358 micro-s per loop
```

Determinants are often employed in linear algebra because they are related to the invertibility of a matrix, and on whether the associated linear system of equations is solvable (a smaller determinant can indicate that some eigenvalues are close to zero, which makes inversion harder). However, to calculate the determinant of extremely large 2D arrays as we will often do in this volume is computationally so demanding that we will do it only for small problems. Let us calculate the determinant of few random matrices, 10×10 and 100×100, for example:

```
In [328]: for i in np.arange(10):
dets[i]=linalg.det(np.random.rand(10,10))
In [329]: dets
Out[329]:
array([-0.00117953, 0.00443444, -0.00628948, 0.00481016, 0.00241629,
-0.02581865, -0.00253544, 0.00121148, 0.00547928, 0.00761418])

In [330]: for i in np.arange(10): dets[i]=linalg.det(np.random.rand(100,100))
In [331]:dets
Out[333]:
array([ -9.89911656e+23,    1.59137713e+26,    2.47655081e+24,
5.69360031e+24,    4.47293786e+25,   -1.56067775e+24,
-3.37434315e+25,    1.61850669e+26,    5.36518712e+23,
2.66219999e+24])
```

Where one observes the dramatic increase in the determinant value, but also how the order of magnitude of the determinant is strongly related to the size of the Matrix. In the graybox it is shown how one can estimate this relationship in depth.

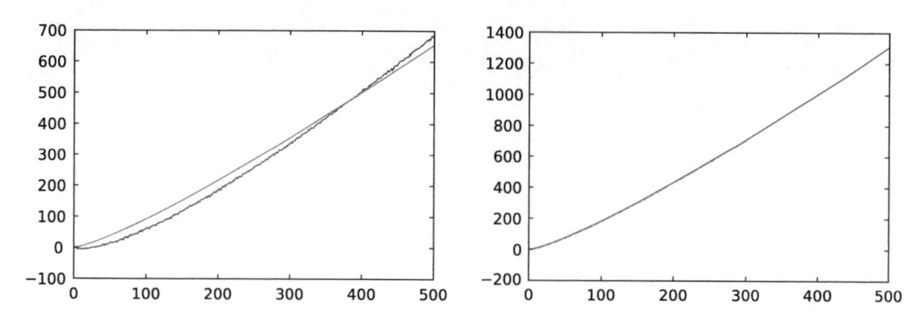

Fig. 3.2 *Left* in log-log plot the *blue line* represents the logarithm of the absolute value of the determinant of a A_n random matrix in function of n where to random values are between uniformly chosen between 0 and 1. The *red line* represents the logarithm of the Stirling style approximation divided by 4. *Right* the same comparison taking a normal distribution for the random matrices versus half of $log(n!)$. Note the striking match!

This an advanced topic, inserted here only for training and enjoyment not necessary to understand the rest of the book.

Random matrices have represented for more than 50 years a fundamental topic of very intense research because of their many applications. Basically for every physical phenomena that can be mathematically formulated in a linear form its statistical properties are expressed by random matrices. Mathematical details can be found in a recent review [44].

One of the most fundamental properties of random matrices is how its determinant scales. The most recent estimations show that a good approximation for the determinant a A_n matrix is $\sqrt{n!}$ for a slowly growing function of n. We can test this approximation using the Stirling formula ($n! \approx \sqrt{(2\pi n)}n^n$) and our calculations of the determinant. Notice how I had to use *np.float128* types in order to estimate properly the huge numbers that emerge. The result of the test below is on the left of Fig. 3.2. On the next figure, the striking comparison of the $\sqrt{n!}$ with the calculation of random matrices using the Normal (Gaussian) Distribution.

```
import numpy as np
import numpy.linalg as linalg
import matplotlib.pyplot as plt

n=500 #maximum matrix size
ns = np.arange(n,dtype=np.float128)+1 #matrix sizes
dets = np.zeros(n, dtype=np.float128) #determinants
facts = np.zeros(n, dtype=np.float128) #sqrt of n!
for i in np.arange(n):
    dets[i]=linalg.det(np.random.rand(ns[i],ns[i]))
    facts[i]=(2*3.14*ns[i])**0.5*(ns[i]/2.71828)**(ns[i])

plt.plot(np.log(abs(dets)))
plt.plot(0.25*np.log(facts)) #1/4 for the uniform distribution
plt.show()
```

3.7 Cython

There are situations in which a simple and straightforward NumPy solution does not exist or can be very difficult to implement. We will see some examples in the last, advanced part. When we are in this situation, it is first of all important to verify first whether this lack of vectorization makes our code sensibly slower. Since this is often the case, a powerful solution has been created by the authors of the parts that compose the Python Scientific Environment, which is called *Cython*.

The most recent updates and documentation for Cython are available at the address `cython.org`). Cython is based on Pyrex, library that allowed to extend Python with C functions. Recently, however *Cython* developers went beyond and created an *extended* language where codes written in Python are optimized and compiled just at the moment of running the Python program. In other words, while the Python language is interpreted, the computer will on-the-fly compile the parts of your code written in Python in special Cython files. Still they will be written in Python, which means that the user does not need to learn a new, more challenging language, such as C. The result is a superior language, only slightly more complex than Python, that combines the best of the two worlds, C and Python [64].

The main addition of Python to Cython are *static variables*. These are variables for which type and space in memory is predefined by the programmer. The advantage of using them is that the Python parser does not need to understand how to interpret them on the fly (one of the most original but computationally most inefficient characteristic of Python) while will straightaway compile and execute the fastest possible way to run the code. In practice, *Cython* is simply Python code with variables defined using *C data types*. Typically it consists of one or few functions that require being compiled with explicitly defined types and called in our code. Since the performance of *Cython* code are nearly equal to an equivalent C-code, but are much easier written, Cython is the definitive tool that makes prototyping in Python the *easy and powerful* programming tool that every scientist dreams.

3.7.1 Cython in iPython

Let us look at some examples of how it works, starting with iPython. I will use IPython 4.0.1. Different versions of IPython can make use of other magic commands.

In iPython first you have to load the Cython library. This can be done with the magic command:

```
In [500]: %load_ext Cython
```

Immediately after, one can use both the magic commands *%cython*, and *%%cython*, the first to run external Cython files, and the second to run a Cython cell. Let us see how we could have implemented the function *setNegativeValuesToZero* that we introduced earlier in the chapter. A first attempt would be to simply rewrite that function, just adding that *n*, *m*, *i* and *l* are integer types:

```
In [501]: %%cython
     ...: def setNegativeValuesToZero(int n,int m,a):
     ...:     cdef int i, j
     ...:     for i in range(n):
     ...:         for j in range(m):
     ...:             if a[i,j]<0:
     ...:                 a[i,j]=0.
```

As one reads on line 2, the type declaration is done simply by using the instruction *cdef*, then followed by the standard static C type declaration. Notice also that I replace *np.arange* with *range*, since this is the compiled version where loops are compiled. Basically, besides the addition of the *int*, i.e., integer, inside the call, this is exactly the same function introduced earlier. One can test its performance with the same random 2000×2000 array finding an improvement of a factor two compared to the pure Python version:

```
In[ 502]: %timeit setNegativeValuesToZero(2000,2000,arr2d)
1 loop, best of 3: 972 ms per loop
```

Can one do better? Yes, but it requires some more work. I show here a version that performs practically exactly the same as the NumPy one:

```
In [503]: %%cython
   ...: cimport numpy as np
   ...: def setNegativeValuesToZero(int n, int m, np.ndarray[double, ndim=2] a):
   ...:     cdef int i, j
   ...:     for i in range(n):
   ...:         for j in range(m):
   ...:             if a[i,j]<0:
   ...:                 a[i,j]=0.
```

As one observes here, it is necessary to add one more line and change one type. The first is the import of numerical python within the Cython framework. This allows to set the type of the *NumPy array a*, which in turn allows Cython to accelerate its process. Once these two little modifications are done, one obtains the striking result:

```
In[504]: %timeit -n 1 setNegativeValuesToZero(2000,2000,arr2d)
1 loop, best of 3: 4.94 ms per loop
```

Which is very close to NumPy, and in this case slightly better.

A good Python programmer, although not *Cython* expert, will need only ten minutes to figure out how to modify a routine to make it ready for Cython, however for the novice programmer to go through the debugging effort that compiled programs require can be quite challenging. For this reason I recommend always using the straightforward *NumPy* libraries whenever necessary, and rely on *Cython* only when really necessary. The many more details that one has to know in order to program *Cython* are available in the excellent recent book *Cython - A Guide for Python Programmers* [40].

3.8 Going Parallel: mpi4py and PETSc4py

In the mid of the years 2000, a revolution has changed modern computing. Until there the calculation power of processors had duplicated about every 1.5 years, as predicted by the Moore's law. This progress, however, has hit a wall and in the past 10 years the speed increase for our computers has been only guaranteed by increasing the number of cores, implying that efficient software needs to be thought again, in terms of parallel software.

Many libraries for efficiently running multicore and multiprocessor software exist, but only few of them have reached a worldwide agreed standard. The Message Passing Interface (MPI) went through several stages of standardization (MPI1, MPI2) and is today the most used system on a wide variety of parallel distributed computers. The MPI standard defines the syntax and semantics for the main scientific programming languages (Fortran, C, or C++). Among open source, the most well-known projects are MpiCH and OpenMPI.

I will review here how to use *mpi4py*, that is a simple but powerful interface that allows to write MPI software natively in Python. This is however not the only option, and other projects exist such as pyMPI, that is also very simple but does not support numerical python arrays, and *pypar*.

mpi4py is presently a translation of standard *MPI-2* bindings for C++ to Python. There are two categories of variables that can be passed from one processor to the other: (i) the standard Python objects, and (ii) the NumPy arrays. Because the memory configuration of the first kind is not compact nor standard, *mpi4py* achieves maximum performance by building binary representations of these objects that are then sent to other processors. Clearly building this binary representation limits its use and can create a serious overhead for very large systems or data. Furthermore this process, called *pickling*, can also require extra memory and it needs to be tested on every machine.

In order to use *mpi4py*, one has to install first MPI independently from the Python installation. Presently, on the windows platform, one has to install OpenMPI through Cygwin. Once this has been done, one can install *mpi4py* in Cygwin, or any other Linux system as well as the OSX Terminal, simply with the command *pip install mpi4py*. If using Anaconda the command *conda install -c anaconda mpi4py=2.0.0* will add *mpi4py* to the Anaconda installation.

Once *mpi4py* is installed one can immediately test the two main techniques mentioned earlier. I show first how to send a list with the picking option (the slow one) from processor one to processor two. It is very straightforward. A list called *data* is sent from processor 0 to processor 1. The Python program knows in which processor it is running by getting the *rank*. If *rank* is 0, it sends the data, if *rank* is 1, it receives it.

```
from mpi4py import MPI

comm = MPI.COMM_WORLD
rank = comm.Get_rank()

if rank == 0:
    data = {'a': 0.1, 'b': 1, 'c':'house', 'd':1.0e100}
    comm.send(data, dest=1, tag=111)
    print('data sent from',rank,' :',data)
elif rank == 1:
    data = comm.recv(source=0, tag=111)
    print('data received from',rank,' :',data)
```

The outcome of this routine looks like this, where I called the file *mpi4py-picking.py* . You can test your machine, it should be quite similar:

```
$ mpiexec -n 2 python mpi4py-picking.py
data sent from 0   : {'c': 'house', 'b': 1, 'd': 1e+100, 'a': 0.1}
data received from 1   : {'d': 1e+100, 'b': 1, 'c': 'house', 'a': 0.1}
```

When working with NumPy arrays, however, arrays are already stored in the memory in the C-format, therefore only the address and size of the memory buffer needs to be passed. This allows MPI implementation with *mpi4py* to reach performance close to C-speed and excellent memory management. We will mainly use the highly optimized case of distributing and processing NumPy arrays in the parallel implementations shown in this book. *mpi4py* is in fact able to recognize their type, but the allocation of the memory space has to done accordingly:

```
if rank == 0:
    data = np.arange(10000, dtype=np.float64)
    comm.Send(data, dest=1, tag=13)
    print('rank:',rank,'sent np.float64 data')

elif rank == 1:
    data = np.empty(10000, dtype=np.float64)
    comm.Recv(data, source=0, tag=13)
    print('rank:',rank,'received np.float64 data')
```

Which gives an outcome like:

```
$ mpirun -np 2 python mpi4py-numpy-arrays.py
rank: 0 sent np.float64 data
rank: 1 received np.float64 data
```

To solve a simple problem, let us consider the setting the negative numbers of a large array to zero, but in a distributed environment, on 16 processors. This is an especially easy case of a so-called *embarrassingly parallel* problem, since one can work on a part of the array without caring about the elaboration of the rest. No passage of information is necessary during elaboration, however this is still an useful example to start to learn how to use MPI.

Here I show a routine that splits the *arr2d* that I created before in *p* equally sized parts, sends them each to a processor, makes the simple elaboration that we need here, and sends them back to the main processor.

```
from mpi4py import MPI
import numpy as np

comm = MPI.COMM_WORLD
size = comm.Get_size()
rank = comm.Get_rank()
```

```
p = size # number of processors
n1,n2=8000,8000
# the array in each processor has a (n1Local,n2) shape
n1Local = np.int(n1/p) # correct if n1 can be devided by p
arr2dLocal = np.empty((n1Local,n2),np.float64)

if rank == 0:
    arr2d = np.random.randn(n1,n2)
    print('non zero components of arr2d:',arr2d[arr2d<0.])
    arr2dLocal = arr2d[0:n1Local,0:n2]
    # send one slice of the array to the other processors
    for i in np.arange(1,p):
       start = i*n1Local; end = (i+1)*n1Local
       comm.Send(arr2d[start:end,0:n2], dest=i, tag=11)
else:
    #receive the slice at each processor (also zero)
    comm.Recv(arr2dLocal, source=0, tag=11)

arr2dLocal[arr2dLocal<0.]=0.

if rank == 0:
    arr2d[0:n1Local,0:n2]=arr2dLocal
    # receive all the slices of the array from each processor
    for i in np.arange(1,p):
       start = i*n1Local; end = (i+1)*n1Local
       comm.Recv(arr2d[start:end,0:n2], source=i, tag=22)
    print('non zero components of arr2d:',arr2d[arr2d<0.])
else:
   #send the back the corrected slice to the the zero processor
   comm.Send(arr2dLocal, dest=0, tag=22)
```

This is a very inefficient way to proceed, using loops and subloops. It also does not scale well on a large number of processors because the root processor sends and receives the data in a preordered sequence, increasing the *time lag*. MPI offers in fact many other commands to make this process more efficient, among them there are in mpi4py *Bcast()*, and *Reduce()* for homogenous sets of data, and *Scatter()* and *Gather()* to split a large memory buffer in equal parts. Alternatives are *Allgather()* and *Alltoall()*. There are also vector variants that allow to communicate different amounts of data to each processor called *Scatterv()*, *Gatherv()*, *Allgatherv()* and *Alltoallv()*.

These and many other subtile aspects go beyond the scope of this introductory textbook, and are available in classic textbooks such as [29, 41]. Some key details on the available strategies to use the standard *mpi4py* procedures are available in the MPI for Python manual [69].

Once mpi4py is installed, it is possible also to use it in combination with iPython, which offers the advantages of the interaction. In order to do so, one needs one more step, which is to start one instance of the controller one for the engine. When working on a single host (for example your laptop) this can be done from the terminal with a unique command *ipcluster start -n 2* where the number indicates the number of engines. I have two processors on my laptop, so I will start with 2, but it works well also with 4, or more. When working on a parallel machine or several machines, the process to start the engines is more complex and its implementation is presently not stable. More details are presently available at the site http://ipyparallel.readthedocs.io/.

Once the engines have started, *iPython* works by importing the associated module and calling a new client. For example,

```
In [600]: import ipyparallel as ipp
In [601]: myClient = ipp.Client()
In [602]: myClient.ids()
Out[602]:  [0,1]
```

where the numbers refer to each engine. One can easily experiment with more engines. It is important that every time the engines start again, the client needs to be instanced *a = ipp.Client()*. The magic command to start parallel Python routines is *%px*. This simple command sends one instruction to all the engines, or typically runs one python routine in every engine.

A more applied but also more advanced and more modern way to use parallel programming is to employ *PETSc* [53]. This is a suite of data structures and routines that aims, in general, at solving all the differential equations that appear in scientific computing. Not only *PETSc* supports MPI, but also CUDA and OpenCL (for programming NVIDIA GPUs), and it has routines that allow hybrid MPI-GPU parallelism. A Python library exists for accessing this set of libraries, called *petsc4py*. Its installation and usage are similar to *mpi4py*. A clear and concise comparison between *mpi4py* and *petsc4py* is also available in [58], together with several key examples.

Overall, discussing about parallel computing can go much further and become much more technical, in particular in relationship to the balance between elaboration and memory access. Good references for the ones who are interested to go more in depth are [61, 62].

3.9 Other Computational Modules

Most of the codes that are illustrated in this book can be accelerated using a variety of other tools. For example, *Numba* offers a speedup of one order of magnitude on many codes where large NumPy arrays are used. It is also extremely easy to use. For example, let us compare the relatively complicated vectorial operation $x^6 - x^3 + \log x$ using Numba with the standard NumPy expression. In iPython *Numba* will be simply called by `@numba.vectorize` and then by a normal python code:

```
In[400]: @numba.vectorize
    ...: def operateWithNumba(x):
    ...:     return x**6-x**3+np.log10(x+1)
```

This can be benchmarked using the magic iPython command `%timeit` and it gives on my laptop:

```
In[401]: y=np.arange(1000000, dtype=float)

In[402]: %timeit -n 1 y**6-y**3+np.log10(y+1)
1 loops, best of 3: 92.6 ms per loop

In[403]: %timeit -n 1 operateWithNumba(y)
1 loops, best of 3: 15.1 ms per loop
```

The advantage of using *Numba* over other tools is its being *Just in Time* (Jit). If we need to repeat the same operation on NumPy arrays many times in a code, Numba can be the simplest way to obtain compiled code performance.

Overall Python offers a huge variety of options when time comes to optimize our code. The key question for student/scientist whose main goal is to perform a research that *uses* a software, and not the development of the software itself, is what is the right amount of time and energy that one should devote to accelerating the code. My suggestion is to dive into these modules only when it is clear that the time spent into developing will pay. In other words the main goal of writing a code in Python is to make our life easier, so it is important to not to complicate it again in order to gain a 30% performance increase, while 3 – 4 times performance increase, and overall orders of magnitude performance increase can substantially change the quality of your research.

Summary

- Our computers are so fast that sometimes we do not perceive the time that it takes for them to perform simple operations. `%timeit` allows us to calculate it precisely.
- *NumPy* (Numerical Python) is the main tool that a modeler needs to master. It allows fast operations on multidimensional arrays, Linear Algebra, Fourier, Tree Structures, and fast mapping for input/output of large datasets.
- Indexing is extraordinary fast in NumPy both when simply slicing the arrays but by using complex Boolean Indexing. Strides play a big role in these performance.
- Vector Products and many more complex Linear Algebra operations are optimized at C-speed in NumPy (because written actually in C).
- When it is too complex or simply impossible to devise a NumPy-based implementation, *Cython* allows writing a Python code that is Just in Time compiled and that performs at C-speed. Cython requires the explicit declaration of all the variables involved, otherwise it is just Python.
- Further speed can be gained using parallel computing. MPI is easily accessible with the *mpi4py* library, many optimized linear algebra implementation are available through *PETSC4py*, and GPU can be programmed using *PyCUDA*. Just in Time compiler of single functions can be done using *Numba*.

Problems

3.1 In Sect. 3.1 we have familiarized with `%timeit`. Test the main operations, addition, multiplication, division, exponential, without using NumPy. By repeating an operation for 1 million of times, check whether they are equivalent or not. And also try both `while i<max` and `for a in range(max)` to understand which ones perform better. Do you understand the results?

3.2 Based on what learned in Sect. 3.2.1, write a program that tests for which small ε a and $a + \varepsilon$ become indistinguishable using *np.float32*, *np.float64* and *np.float128* numbers.

3.3 Using the techniques learned in Sect. 3.2.2, compare how standard unary *abs*, *log* operations in Python perform compared to NumPy broadcasting. The two operations are very different, the first requiring a minimum computational cost, the second more intense. Do you observe a difference in performance gain? Why?

3.4 Again, based on the examples in Sect. 3.2.2, test Binary functions such as *power* and *maximum* for arrays of one thousands, one million and if you computer allows it, one billion float numbers. Is the performance gain constant?

3.5 By modifying the procedure shown in Sect. 3.5 to create a regular mesh with outer products, create a mesh that is not regular, with elements at the center two times smaller in every direction than the elements at the edges. There are several equivalent strategies.

3.6 As in Sect. 3.6, calculate the determinant of 100 random 10×10, 30×30, 50×50 and 100×100 matrices. Observe how the variability of the values of the determinant is large for 30×30 and is instead remarkably stable for very large matrices. One can also test several random distributions (without a power-law tail) and observe the remarkably similar determinant values.

Part II
Second Part: Mechanics

Here it is shown how to solve basic problems of Mechanics. The concepts used in this part are the one introduced during a general undergraduate physics course, where, however, only solved using analytical methods are used. Here I show how that same problems can be simulated using Python. By doing so, the students without experience in programming can quickly develop the necessary *Forma Mentis* for developing a geodynamic simulation software.

Once the concepts are introduced, the methods are applied to simple cases related to the geosciences. For example, the general case of rain on Earth as well as on other planets. This allows showing how fundamental physical concepts such as the conservation of momentum and of energy can be used to check the consistency and correctness of the implementation of the simulation itself.

It is also shown how simulating the dynamics of a large number of objects allows to understand statistical properties of a certain system. This technique, called *Monte Carlo Method* is very general and is employed to numerically characterize many physical systems. Here it is applied to the fall of volcanic lapilii (pyroclasts) around the volcano edifice during an eruption.

Finally continuum mechanics is introduced in a non-systematic way, showing some key examples that help to understand fundamental concepts. In particular it is emphasized how Stokes Flow solutions differ in 2D and 3D and how this has to be taken into account when modelling geodynamics.

Chapter 4
Mechanics I: Kinematics

"Acceleration is finite, I think according to some laws of physics."

— Terry Riley

Abstract It is shown here how to use Finite Differences to calculate displacement, velocity, and acceleration using a discretized approximation, that in a general undergraduate Physics course would be obtained by using derivatives and integrals. Both ways are illustrated (i) the calculation of the kinematic quantities from a known trajectory, and (ii) the calculation of the trajectory by knowing the acceleration field in the space. The possible errors that can be made are illustrated in detail, in particular how a small error in calculating a trajectory can amplify when solving for a long trajectory.

In this and in the following chapter, we will consider very simple problems that are normally encountered in a general Physics course, relative to Kinematic and Newtonian mechanics. We will focus on how all these problems can be solved by using a numerical approach only, without any analytical calculation. For many of the cases considered here, it is simply an overkill to use a computer to solve them, however because of their simplicity they are an excellent way to introduce an inexpert reader to new numerical techniques.

The strategies that we use here will be useful in the rest of the book to understand how to calculate the motion of particles in the Particles in Cell method. They will give us the opportunity to simply introduce Eulerian (mesh based) and Lagrangian (particle based) approach that are essential when solving problem in the more complex deformable media.

In this and the next chapter, we will initially assume that every object is rigid, and introduce only generically the concept of strain, i.e., the quantification of the deformation of the objects themselves. In fact while the velocity and the acceleration are the derivative of the displacement in time, the strain is the derivative of the displacement in space, and the strain rate both the derivative of the displacement in space and in time. With them, combined with the diffusion process, we can describe every process that happens in the solid interiors of a planet (the liquid core, as well

G. Morra, *Pythonic Geodynamics*, Lecture Notes in Earth System Sciences,
DOI 10.1007/978-3-319-55682-6_4

as stars interiors, require also the treatment of the electromagnetic equations that go beyond the scope of this introductory text).

Overall the goal of this chapter is to start to familiarize with programming in Python and to understand the importance of the concepts of momentum and energy. In all the problems in this chapters only the time is discretized, meaning that the solutions are calculated for a time t and then for the next time step at $t + \Delta t$, where Δt can be constant (regular discretization), or vary. The discretization of the physical space, instead, will be introduced in Chap. 7 together with the equations that control the deformation of a stressed body.

4.1 Computation of Velocity and Acceleration

Kinematic quantities are ultimately defined in function of some time derivative of the position of an object, in cartesian coordinates $\mathbf{x}$ or angular coordinate $\boldsymbol{\theta}$. Let us learn how to express these equations at discrete times t_i.

Since we describe all variables in function of time, we have to define a time array first. I recommend to use here IPython to reproduce step by step the shown examples, and play with them. However, for readability I will not show *In[]* and *Out[]*.

Let us define a regular array $t_0, t_1, t_2, ...$ where $t_i - t_{i-1} = \Delta t$ are non-dimensional time intervals. If we initially divide the time range in 20 intervals assume time nondimensional between 0 and 1, we can write:

```
import numpy as np
tmax = 1.0
tmin = 0.0
intervals = 20
dt =(tmax-tmin) / intervals
nt = intervals + 1
time = np.arange(nt) * dt
```

Observe that *nt* is a unity greater than *intervals* in order to that 0 and 1 are both in the array. *dt* is 0.05, i.e., Δt. Once time is defined we can create arrays for position, velocity and acceleration. Initially the position can be set to zero:

```
x=np.zeros(nt)
```

Let us first consider the easiest case, the one in which we know the position of the object in time and we want to extract velocity and acceleration. Harmonic motion is a classical example: if the position of the object oscillates (like the horizontal position of a sphere attached to a pendulum) the position will follow a law like $x = sin(2\pi t)$, where the initial position was set to *zero* and the time necessary for an entire oscillation is *one*. In terms of a Python array:

```
x[0:nt]=np.sin(2*np.pi*time[0:nt])
```

We could have also defined immediately `x=sin(2*np.pi*time)`, which would have implicitly allocated the memory, but I am going step by step so that it is clear how we are managing memory. x is a new array whose values are stored elsewhere from *time*. Remember that the instruction `x=time` instead would have not allocated a new memory chunk (see Sect. 3.3), as one can easily verify with the instruction `x=time;x[10]=0;print(time[8:13])`.

Velocity and acceleration are defined as the the first and second derivative of position in time, respectively, therefore they can be approximated in function of the time discretization:

$$v(t) = \frac{dx(t)}{dt} \approx \frac{x(t_i) - x(t_{i-1})}{t_i - t_{i-1}} = \frac{x_i - x_{i-1}}{\Delta t}$$

where we have assumed that $\Delta t = t_{i+1} - t_i$ is independent from i. While intuitively correct, the above approximation is however biased because oriented *backward*, because approximating the derivative at time t_i at its left side. A symmetric *forward* formulation would be equally logical:

$$\frac{dx(t)}{dt} \approx \frac{dx}{dt}(t_i) = \frac{x(t_{i+1}) - x(t_i)}{t_{i+1} - t_i} = \frac{x_{i+1} - x_i}{\Delta t}$$

but it would be less practical since in many problems we do not know the future position. The most natural definition of derivative is the *centered* one, that is just the arithmetic average of the two:

$$\frac{dx(t)}{dt} \approx \frac{1}{2}\left(\frac{x(t_{i+1}) - x(t_i)}{t_{i+1} - t_i} + \frac{x(t_i) - x(t_{i-1})}{t_i - t_{i-1}}\right) = \frac{x_{i+1} - x_{i-1}}{2\Delta t}$$

Although not sided *backward* or *forward*, this definition has the odd property of not containing the term x_i, i.e., the position at the time in which we want calculate the velocity. This characteristic produces *dispersion*, i.e., the derivative is slightly more flat than the other approximation. In other words the information on the behavior of the derivative functions is diffused laterally.

The acceleration can be either calculated as the first derivative of the velocity in time or as the second derivative of position in time. The most compact and less diffusive formulation to obtain acceleration from position is obtained by combining *backward* and *forward* formulations:

$$a(t) = \frac{d^2x(t)}{dt^2} = \frac{x(t_{i+1}) - 2x(t_i) + x(t_{i-1})}{(t_{i+1} - t_{i-1})^2} = \frac{x_{i+1} - 2x_i + x_{i-1}}{(\Delta t)^2}$$

We also immediately notice that this formulation of the second derivative is naturally symmetric around i. This is a result of the fact that the *backward* first derivative is symmetric around $i + 1/2$ and that the *forward* first derivative is symmetric around $i - 1/2$. This is also a first insight on a fundamental property of derivatives on a discrete mesh, that they are naturally symmetric and compact in a grid created by points halfway between the nodes of the initial mesh, often called *staggered* mesh.

As a consequence even derivatives are naturally best calculated on the original mesh, while uneven derivatives on the staggered one. We will see that in 2D and 3D this may create complicated formulations both when solving equations on a grid, using finite differences, and when projecting on a grid solutions obtained with other techniques. It also implies that when possible it is much better to solve problems using even derivatives since an optimal implementation can be more easily implemented.

Let us calculate the first derivative using the *forward*, *backward* and *centered* definitions. *Forward* and *backward* derivatives occupy an array that is long as the number of *intervals*, i.e., one element shorter than *position*:

```
dxdtForward = np.zeros(nt)
dxdtBackward = np.zeros(nt)
dxdtCentered = np.zeros(nt)
```

We can now calculate the *forward* and *backward* derivatives. As we observed in the past chapter, this operation is much faster if instead of a loop operators such as *for* or *while* we use the properties NumPy arrays:

```
dxdtForward[0:nt-1] = (x[1:nt]-x[0:nt-1])/dt
dxdtBackward[1:nt] = (x[1:nt]-x[0:nt-1])/dt
```

Using this definition we also observe that backward and forward derivatives in time are the same, it just depends on whether *dxdt* is associated to *time[1:nt]*. Thinking at the concept of the staggered grid that we just introduced, it means that either we are shifting the grid backward or forward in time.

Different is the case for the centered derivative. This is in fact not immediately defined at the edges, but only for $nt - 2$ points. We can use again the fast array operations to define it as:

```
dxdtCentered[1:nt-1] = 0.5 * (dxdtForward[1:nt-1] +
dxdtBackward[1:nt-1])
```

Notice that the allocation of the NumPy array was automatically performed by Python, however *dxdtCentered* will be 19 elements long. We could have equally defined it as *dxdtCentered = (x[2:nt]-x[0:nt-2])/(2*dt)*, obtaining the same result.

Let us now plot the *position* in function of time using the Matplotlib tools that we introduced before:

```
import matplotlib.pyplot as plt
```

Using the options *color* and *label* I have can set colour and a name for the figure legend for each curve. The legend will be displayed only when explicitly stated with the command *plt.legend()*.

```
plt.plot(time, x, color='b', label='position')
plt.plot(time[0:nt-1], dxdtForward[0:nt-1], color='r', label='Forward')
plt.plot(time[1:nt], dxdtBackward[1:nt], color='g', label='Backward')
plt.plot(time[1:nt-1], dxdtCentered[1:nt-1], color='c', label='Centered')
plt.legend(loc = 'upper center')
plt.show()
```

When plotting the three velocities we observe a skew between them in Fig. 4.1. It is important that the difference between *forward* and *backward* depends on the chosen subset of the time array, right (*time[1:nt]*), and left (*time[0:nt-1]*) respectively. In other words the difference is in the background grid.

These differences may not seem important here, but we will see that this small effect adds up. Clearly by increasing the resolution, this offset will fade. However to increase the number of details is only a good solutions at low dimensions, like 1D or 2D. In 1D in effect one can effortlessly go up to 100,000 points. However, we want in the future to develop models at higher dimensions. In 3 dimensions a $100 \times 100 \times 100$ mesh requires already one million points. Increasing resolution only up to $500 \times 500 \times 500$ already demands solving our equations on over 100 million points, which is not an easy task and may require distribution of memory resources and computing time. For these reasons carefully planning a reliable discretization scheme for application in the physical world is a priority.

Using the same strategy, we can calculate the acceleration using the simplest stencil:

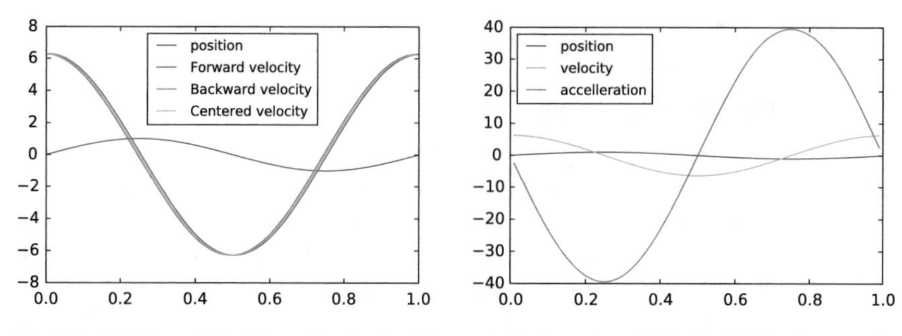

Fig. 4.1 *Left* three definitions of velocity: Forward, Backward, and Centered. The time on the x-axis is discretized with 20 points. The three velocities almost overlap, but a shift is visible. *Right* Position, Centered Velocity, and Acceleration. The maximum values of the velocity is equal to $2 * \pi$ times the position, and so is maximum acceleration versus maximum velocity. The last value on the *right* of the velocity and the edge values of the acceleration are missing

```
d2xdt2 = ( x[0:nt-2] - 2*x[1:nt-1] + x[2:nt] ) / dt**2
```

The array containing the second derivative *dx2dt2* has the same length of *dxdtCentered*. We can plot now position, velocity, and acceleration together:

```
plt.plot(time,x,color='b',label='position')
plt.plot(time[1:nt-1],dxdtCentered,color='c',label='velocity')
plt.plot(time[1:nt-1],d2xdt2,'g',label='acceleration')
plt.legend(loc='upper left')
plt.show()
```

This plot is shown in Fig. 4.1. The peak velocity and acceleration are at a factor of 2π one from the other, which is coherent with the definition that we adopted. We also observe that the last point at the edges is missing, due to the impossibility to calculate the derivative with this formulation. If this value must be calculated optimal approximations for the first and second derivatives exist at the left side:

$$\frac{dx(t)}{dt} \approx \frac{-3x_i + 4x_{i+1} - x_{i+2}}{2\Delta t}$$
$$\frac{dx^2(t)}{dt^2} \approx \frac{x_i - 2x_{i+1} + x_{i+2}}{\Delta t^2}$$

And symmetrically at the other side. Approximations like these ones are based on the minimization of the truncation error of series expansions. The interest reader in how to approximate finite differences close to edges can find more details in many textbooks, e.g., [23, 34].

4.2 Integrate Acceleration

Integration is the inverse functional (it is not a function) of derivative. When we know the acceleration we can obtain the velocity with a first integration and then the position with a second derivative. Let us consider the opposite problem of Sect. 4.1: an object is thrown vertically in the air with the initial velocity v_0 from the initial position $y_0 = 0$ and under the acceleration of gravity $a = -g = -10\,\mathrm{m/s^2}$. Since we now know the acceleration and the initial conditions, we know that we can find the entire kinematic solution, i.e., position and velocity versus time.

I assume that *NumPy* and *matplotlib.pyplot* are imported, and I will also use the same values for *nt*, *dt*, *intervals*, *tmin* and *tmax*. Let's initially define an array for time.

```
time = np.arange(nt) * dt
```

Although in this case the acceleration is constant, it is practical to set its value in an array, as earlier. This will also allow us to extend this technique to a nonconstant acceleration (e.g., a rotating system). Due to the finite difference approximation, we assume that its length is of two elements less than *nt*. We expect that at every integration the size of the arrays (velocity and position), will increase, opposite to the derivative operator.

```
acc = np.ones(nt) * (-10.0) #m/s2
```

Let us now create an empty vector that contains the velocity and let us initiate the value at time $t_0 = 0$. Here I choose $v_0 = 5.0\,\mathrm{m/s}$:

```
vel = np.zeros(nt)
vel[0]=5.0   #m/s
```

We can now perform the integration and calculate the velocity with the inverse operation respect to numerical derivation. In general, we cannot just calculate *vel* as a vectorized operation as we did before, because the integral is the sum of time dependent increments (given by the acceleration), and, if we want to develop a general method, we have to assume that we do not know the values of the acceleration in time.

Clearly, because the gravity is constant, we could calculate the solution in one line, and we can calculate the analytical solution, but we will use it only as a bench-mark. In other words we keep our solution strategy general by calculating velocity and position at every time increment. This allows us to apply this strategy to real cases, such as planetary orbits where the acceleration depends on the position, i.e., on the integration history. It is known, for example, that in the general case the long-term analytical prediction of the motion of three or more bodies is intractable, which implies the use of reliable numerical methods. As a conclusion, we use now a *for* loop for the forward time integration, and so we will do in the rest of the book. In most cases this will be the only *for* loop in the entire simulation program.

Starting from the first value v_o to the last one. For the specific case of constant acceleration it is not relevant whether our integration is *forward* or *backward*, so we can simply write:

```
for it in np.arange(nt-1):
    vel[it+1] = vel[it]+acc[it]*dt
```

Let us repeat the operation for the position. Note that the array's length increases at each integration. In this case the velocity is not constant, and the small difference between using the past or the next velocity value adds up when calculating the position:

```
pos = np.zeros(nt)
pos[0] = 0.0 #m
for it in arange(nt-1):
    pos[it+1] = pos[it] + vel[it]*dt
```

In analogy with the past section, where I used the past velocity *vel[it]* to integrate the position, I am here assuming a *forward* approximation. Calculating the average between two velocities in two consecutive points will allow us to have a much better estimate of the average velocity for every interval, and therefore avoid any kind of bias. This can be implemented by creating a vector that contains this average *velocity* and then use it for the integration:

```
posAve = np.zeros(nt)
velAverage = 0.5 * ( vel[0:nt-1] + vel[1:nt])
for it in np.arange(nt-1):
   posAve[it+1] = posAve[it] + velAverage[it]*dt
```

Using the plotting tools learned in the past section one can now visualize the different results:

```
plt.plot(time,pos,color='b',label='with backward velocity')
plt.plot(time,-0.5*10.*time**2+5.0*time,linewidth=4,
  ↪ color='g',label='analytical solution')
plt.plot(time,posAve,color='c',linewidth=3,label='with ave velocity')
plt.legend(loc='lower center')
plt.show()
```

In Fig. 4.2, I show the difference between the two techniques of integration and the analytical solution.

Since we have chosen an initial velocity $v_0 = 5.0\,\mathrm{m/s}$ and assuming a gravity $g = 1.00\,\mathrm{m/s^2}$ the analytical solution is simply $x = x_0 + v_0 t + (1/2)at^2 = 5.0t - (1/2)10.0t^2$, we expect the trajectory to be completely symmetric and the object to return to the initial position after one second $x(1s) = 5.0 - (1/2)10.0 = 0\ m$ and $v(1s) = v_0 + at = 5.0 - 10.0 = -5.0\,\mathrm{m/s}$. This analytical solution allows us to test our approximations. What we find is that for the *central* approximation the numerical solution and the analytical one are nearly indistinguishable. On the contrary, for the *forward* solution the distance between the two curves increases with time and even reaches a different maximum.

4.3 Projectile Trajectory

When we extend the procedure of Sect. 4.2 to two dimensions we obtain a trajectory in space. The independent variable is always *time*, but we solve now for *x(time)* and *y(time)* simultaneously. The solution represents the trajectory of the body. Assuming a constant gravity acceleration vector **g** oriented toward the negative y direction, the trajectory will depend only on the initial velocity **v**.

Let us assume that *NumPy* and *Matplotlib.pyplot* are loaded and that *nt*, *dt*, *intervals*, *tmin* and *tmax* have the same values assumed in the past section. We have now to find velocity and position in the x and in the y direction. If v_0 is the initial speed and θ is the departure angle, then $v_{x0} = v_0 cos(\theta)$ and $v_{y0} = v_0 sin(\theta)$:

```
theta=np.pi/6
v0=10.0 #m/s
gx=0.0; gy=-10.0 #m/s2
ax = np.ones(nt) * gx; ay = np.ones(nt) * gy
vx = np.zeros(nt); vy = np.zeros(nt)
vx[0]=v0*np.cos(theta)
vy[0]=v0*np.sin(theta)
```

We integrate the velocity field as before, but both in the x and y direction:

```
for it in np.arange(nt-1):
    vx[it+1] = vx[it]+ax[it]*dt
    vy[it+1] = vy[it]+ay[it]*dt
```

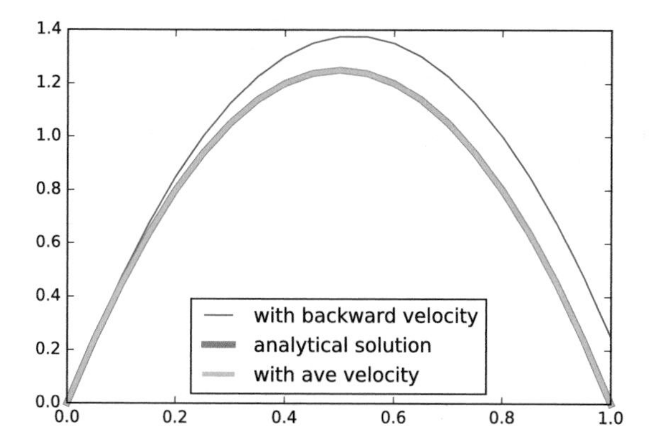

Fig. 4.2 Position vs time of an the object launched at the initial velocity of $v_0 = +5\,\mathrm{m/s}$. Two techniques to integrate the position are employed, and the result compared with the analytical solution. One integration is based on a *forward* approximation in which the velocity for the entire time interval is the initial one. The second assumes the mid interval value. The plot shows that the mid value is much superior. It is in fact so superior that the numerical trajectory is indistinguishable from the analytical solution

We can now calculate the trajectory px and py vs time. As in Sect. 4.2 we use the *central* approximation:

```
px = np.zeros(nt); py = np.zeros(nt)
vxAverage = 0.5 * ( vx[0:nt-2] + vx[1:nt-1])
vyAverage = 0.5 * ( vy[0:nt-2] + vy[1:nt-1])
for it in np.arange(nt-1):
    px[it+1] = px[it] + vxAverage[it]*dt
    py[it+1] = py[it] + vyAverage[it]*dt
```

Which can also be benchmarked vs. the analytical solution $x = v_{0x}t$ and $y = v_{0y}t - \frac{1}{2}gt^2$. The comparison between the numerical solution and the analytical one is shown in Fig. 4.3 for an object launched with an angle of 30° (observe the different x and y scales). The trajectory is now plot as px vs vy and not anymore in function of time:

```
pxBench=v0*np.cos(theta)*time
pyBench=v0*np.sin(theta)*time-0.5*10.0*time**2
plt.xlabel('x coordinate');plt.ylabel('y coordinate');
plt.plot(pxBench, pyBench, color='r', label='Benchmark')
plt.plot(px, py, color='b', label='py vs px')
```

If you are programming this example in *IPython*, the parameters can be changed (e.g., the angle and initial velocity) and the solution recalculated just using the magic command *%rerun*.

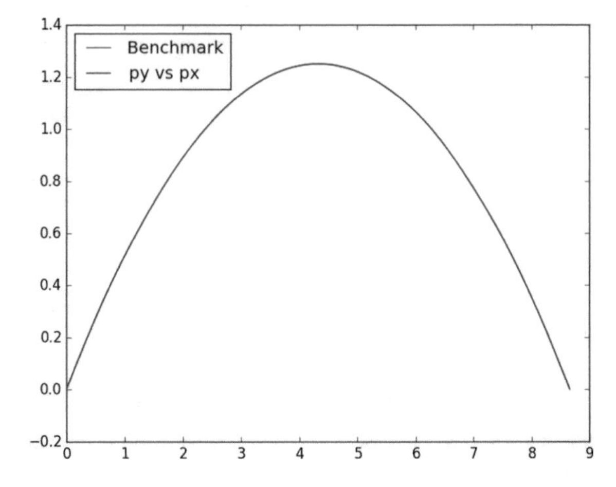

Fig. 4.3 Trajectory (x vs y) of a an the object launched at the initial velocity of 10 m/s at an angle of 30°. The two trajectories display the analytical solution vs. the numerical one. Integration is performed independently in the x and y direction. The two solutions perfectly overlap

4.4 Circular Motion

An interesting way to check the quality of our time integration technique is to use an acceleration that varies with space and in time. In this case, we will have to pay more attention to the strategy chosen to integrate the acceleration.

Let us consider an object in a circular orbit around the Earth. This trajectory will be stable if the centripetal acceleration v^2/r is equal to the gravity acceleration g, i.e., if $v = \sqrt{gr}$. Let us write a program that calculates exactly this orbit and let us use it to calculate other trajectories. For simplicity, we assume a total time $t_{total} = 2\pi$:

```
tmin = 0.0; tmax = 2*np.pi
intervals = 100
dt = (tmax-tmin) / intervals
nt = intervals + 1
time = np.arange(nt) * dt
```

If the trajectory is circular then x and y can be immediately calculated as

```
x = np.sin(time); y = np.cos(time)
plt.plot(x,y,'b',label='position')
```

as well as velocity and acceleration:

```
dxdt=(x[1:nt]-x[0:nt-1])/dt
dydt=(y[1:nt]-y[0:nt-1])/dt
d2xdt2=(x[0:nt-2]-2*x[1:nt-1]+x[2:nt])/dt**2
d2ydt2=(y[0:nt-2]-2*y[1:nt-1]+y[2:nt])/dt**2
```

In two dimensions, it might be interesting to plot velocities and acceleration as vectors. This is possible by using the *Matplotlib* function *plt.arrow*. Let us plot some velocity and acceleration vectors, for example, whose result is shown in Fig. 4.4:

```
numArrows = 20
aLen = 0.1 # length of one arrow
for it in np.arange(0,nt-1,nt/numArrows):
    plt.arrow(x[it+1], y[it+1], dxdt[it]*aLen, dydt[it]*aLen,fc='g')
    plt.arrow(x[it+1], y[it+1], d2xdt2[it]*aLen, d2ydt2[it]*aLen,fc='r')
```

The above exercise can be inverted, and the trajectory extrapolated by a centered acceleration around one or many planets. Exercise 4.2 proposes to write a program that solves this case for one planet. For this specific case, it is possible to find these informations either numerically or analytically. This problem can also be solved analytically for one or two planets, but it becomes intractable for three or more planets that exert a non-negligible force each other. For this case numerical solutions are essential and complement analytical treatment of a potential approach, and we will look at them in the next chapter.

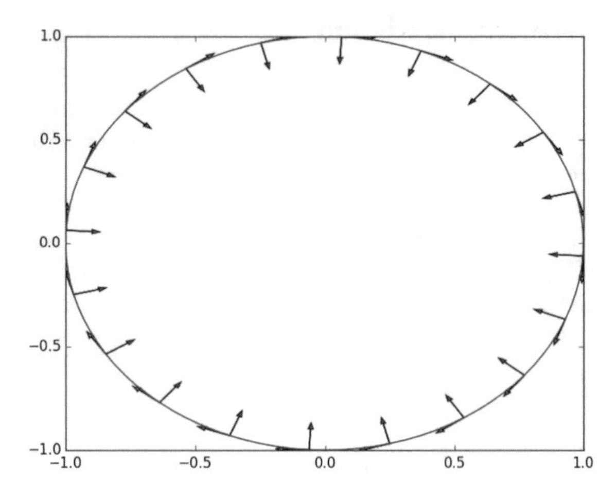

Fig. 4.4 Velocity and Acceleration of an object in a circular motion. The *arrows* are scaled as 10% of the of the magnitude of velocity (*green*) and acceleration (*red*)

Summary

- At the first-order velocity and acceleration can be simply formulated for a discrete time using the finite-difference approximation.
- Finite difference can be centered, forward, backward, and have high order (not necessarily better) more complex formulations.
- Reversely, velocity and acceleration can be integrated in order to reconstruct trajectories. Small changes in the integration technique can cause large errors.
- Circular Motion allows benchmarking a formulation by detecting errors after many rounds. The calculation of centripetal acceleration is a strong test of the quality of the implementation.

Problems

4.1 Using the strategy introduced in Sect. 4.1, calculate discretized formulations of second derivatives using combinations of *forward*, *backward*, and *centered* first derivative approximations. Which ones are symmetric? Which one is the most dispersive? Which ones are symmetric around a different point than t_i? Higher order discrete formulations also exist: any idea of how they are formulated?

4.2 Compute the orbital trajectory of an object assuming that the acceleration is equal to $a_c = GM/r^2$ where r is the distance between the planet center and the object, M is the mass of the planet, and G is the gravitational constant $6.674 \times 10^{11}\mathrm{Nm}^2/\mathrm{kg}^2$. Assuming that the direction of the initial velocity is perpendicular to the vector $\mathbf{r}$ that connects the center of the planet to the body, the trajectory will be either spiraling toward the planet, or spiraling out. What is the critical value that separates the two attractors? What happens if the initial direction is not perpendicular to $\mathbf{r}$?

4.3 Every child makes the experience of leaving an object at the top of a sphere and letting it sliding down until, at a certain height, the object leaves the sphere and falls down following a parabolic trajectory. Assuming that the sphere is frictionless, at which height will fly off the sphere? Numerically solving this problem will show a general simple result.

4.4 A nontrivial problem in physics where numerical approaches are at hand was proposed by Feynman in his famous lectures. In this example, a boat has to run across a channel. In this channel water is streaming at a certain constant speed. The dock on the other side to reach is exactly in front to the starting one. There are generically two strategies to reach the other side:

A. Crabbing. Maneuver the boat steadily pointing with its nose at a certain angle upstream. In this way the boat maintains a fixed orientation and crabs in a straight line across the channel.
B. Pointing. Maneuver the boat in order that its nose points directly at the dock that one wishes to reach.

The exercise is to write a software that simulates the two methods and calculates which one gets the boat to the dock faster. Analytical solutions exist for the case in which the boat runs at a constant speed, although not they are not immediate. A numerical approach, however, allows testing more complex water flow patterns. Can you find a better solution replacing the condition of constant speed with another one with the same average speed?

Chapter 5
Mechanics II: Newtonian Dynamics

When asked about which scientist he'd like to meet, Neil deGrasse Tyson said, "Isaac Newton. No question about it. The smartest person ever to walk the face of this earth. The man was connected to the universe in spooky ways. He discovered the laws of motion, the laws of gravity, the laws of optics. Then he turned 26."

Abstract It is introduced here the concept of force and shown how Newtonian problems, meaning that can be solved from the equation $F = ma$, can be discretized and solved using the Finite Difference method. Applications are (i) the addition of the air Drag to the gravity force on a falling rain drop; (ii) the trajectory of lapils from an explosive volcano, and (iii) the rotation of a Gyroscope (as well as the Earth). For all these systems the Kinematic, Potential, and Dissipated energies are calculated. It is shown how by monitoring their values it is possible to assess the correctness of the numerical result and to find the smallest departure from the exact solution.

We leave now the kinematic world of Chap. 4, where we looked at how to compute velocity and acceleration knowing position versus time and vice versa. In this chapter we will consider how to calculate the forces involved, the linear and angular momentums, and extend the treatment to potential, kinematic, and dissipative energies.

5.1 Analytical Solutions for 1D Dynamics

There are few things more enjoyable than a light rain during a very hot summer. We rarely stop and think that this same rain could be dangerous elsewhere than our planet. In fact few factors make rain such an innocuous phenomena. One of them

G. Morra, *Pythonic Geodynamics*, Lecture Notes in Earth System Sciences,
DOI 10.1007/978-3-319-55682-6_5

is the friction of the air. What would happen if the air would not slow rain? In this case, the only force driving the rain drop would be the force of gravity. After falling from a typical cloud's height (e.g., $h = 2000$ m), its speed at the arrival to the ground would be determined by the equivalence between the potential mechanic energy that would have been turned into kinetic energy:

$$gmh = \frac{1}{2}mv^2 \tag{5.1}$$

Eliminating m one finds $v = \sqrt{2gh}$, that is 200 m/s, i.e., 720 km/h, more than half of the speed of sound in air and certainly enough to kill a person! This is not only the velocity of a drop of water, but the velocity at which any object of any mass would have when falling without the resistance of air. On Mars the force of gravity is less, still the atmosphere is so rarefied (the density is 1% of the one of the Earth) that an hypothetical rain from 2000 meters high, traveling at a speed next the one of a shotgun bullet, could be deadly (not that one would survive without protection on its surface anyway! The low pressure would let our blood into a boiling state, low temperature would freeze us, the dust on the surface is toxic and unfiltered solar radiation would kill us in few months).

Let us calculate analytically the real terminal velocity of the droplets on Earth. This will allow us to have a benchmark to setup our numerical solver. We first have to solve the momentum equation. For a rigid, non-deformable object, linear and angular momentum are defined by the product of the inertia of an object (*mass m* for the linear momentum, and *rotational inertia I* for the angular momentum) and the velocity (linear $\mathbf{v}$ and angular ω:

$$\begin{aligned} \mathbf{p} &= m\mathbf{v} \\ \mathbf{f} &= \frac{d\mathbf{p}}{dt} = m\frac{d\mathbf{v}}{dt} = m\mathbf{a} \end{aligned} \tag{5.2}$$

where $\mathbf{f}$ is the force applied to an object and $\mathbf{p}$ the linear momentum of its object. If you do not know linear momentum, intuitively it can be imagined as the *tendency* of a body to keep its present motion in that direction and at its present speed. Formally, the velocity is defined as the time derivative of the position of the center of mass of an object $\mathbf{v} = d\mathbf{x}/dt$.

To find the terminal velocity of the droplet, we have to estimate the drag force of the air. Our physics textbook tells us that the drag exerted by a fluid to a moving object a speed v can be at the first-order approximated by $D = \frac{1}{2}C\rho_{fluid}Av^2$, where A is the area of the section transversal to the direction of motion and C a 'shape constant'. Droplets are not rigid, but we can roughly approximate their shape to a sphere and assume that the surface tension is strong enough so that they behave like a solid. By assuming standard values, i.e., radius of the water droplet ($r = 1.5$ mm), water density ($\rho_{water} = 1000\,\mathrm{kg/m^3}$), constant for a water droplet ($C = 0.6$), air density ($\rho_{air} = 1.2\,\text{kg/m}^3$), we can express the momentum equation at equilibrium setting Eq. 5.2 to 0: $d\mathbf{p}/dt = \mathbf{f} = 0$ for our case:

$$gm - \frac{1}{2}C\rho_{fluid}Av^2 = 0 \tag{5.3}$$

which gives us the value of: $v = 7.45\,\text{m/s}$, about $27\,\text{km/h}$. This is the speed of droplet when it reaches the ground.

Now that we have understood the asymptotic terms, we can investigate the dynamics using a numerical approach, and learn how to calculate a trajectory and its dynamics using *iPython*.

5.1.1 1-D Dynamics

First let us set the parameter values:

```
radius = 1.5e-3 #m
densityWater = 1000 #kg/m3
mass = 4.0/3.0*np.pi*radius**3*densityWater #kg
Const = 0.6
densityAir = 1.2 #kg/m3
Area = np.pi*radius**2 #m2
```

The acceleration can be calculated and integrated by using the same approach shown in Chap. 4. The only complication is that the acceleration depends on the velocity (as the drag does), therefore, it has to be recalculated for every time step. This can be implemented using two arrays, one for the gravity acceleration, the other for the acceleration due to the drag.

```
tmax = 5.0; tmin = 0.0
intervals = 100; dt = (tmax-tmin) / intervals
nt = intervals + 1; time = np.arange(nt) * dt
accGravity = np.ones(nt-2) * (-10.0) #m/s2
accDrag = np.zeros(nt-2) #m/s2
```

By initializing the velocity at the positive value $v_0 = 5.0\,\text{m/s}$, we can calculate the new value at every step, each time recalculating the drag. The direction of the drag force is always opposite to the direction of motion, therefore it has to be multiplied by $v/|v|$:

```
vy = np. zeros(nt-1)
vy[0]=5.0 #m/s
for it in np.arange(nt-2):
   accDrag[it] =  0.5*Const*densityAir*Area*vy[it]**2/mass
   if accDrag[it]>0:
```

```
    accDrag[it] *= -vy[it]/np.abs(vy[it])
  vy[it+1] = vy[it] + (accGravity[it]+accDrag[it]) * dt
```

The solver is not vectorized here because the acceleration depends on the velocity itself. There is in effect a way to vectorize this solution, and we will see in Sect. 5.1.2 how to do it. With the same procedure shown in Sect. 4.3 the 1D trajectory can be calculated and displayed. Since the droplet was initially flying in a positive direction, it will slow down due to both gravity and drag, it will then fall down accelerated by gravity and again be slowed down by the air drag. The resulting acceleration, velocity and positions are shown in Fig. 5.1.

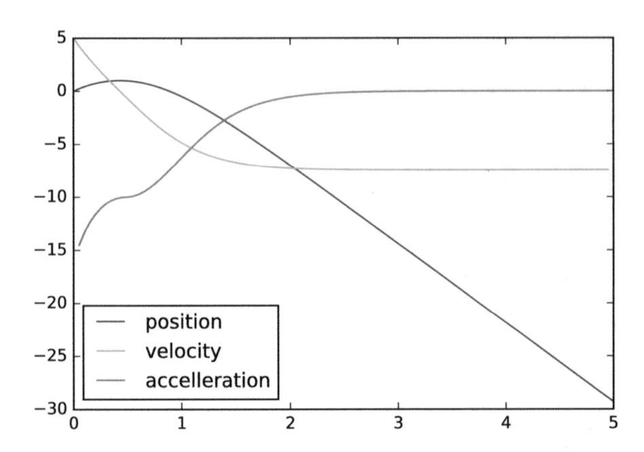

Fig. 5.1 1D time evolution of position (y vs. time), velocity (vy vs. time), and acceleration (a vs. time) of a drop launched towards the sky at the initial velocity of 5 m/s. The flex done by the acceleration is due to the combination of gravity acceleration and drag force

It is interesting to observe how the acceleration changes when the initial velocity is pointing down, up, or is zero. If initially is pointing upwards, one obtains a *plateau* in the solution of the acceleration. This happens because the drag becomes zero when the droplet reaches the highest point and the velocity (and therefore the drag) becomes zero.

5.1.2 2D Dynamics

The above solution can be extended in a straightforward manner to 2D. Since the drag always opposes the speed, here its vector has two components that depend on the direction in which the drop moves. Using the same material constants as in the 1D case, we need only the initial angle and velocity.

```
theta=np.pi/3.; vel=5.0 #m/s
gx=0.0; gy=-10.0 #m/s2
```

In this simulation the variables and array are just a 2-dimensional extension of the past section:

```
tmax = 5.0; tmin = 0.0
intervals = 100; dt = (tmax-tmin) / intervals
nt = intervals + 1; time = np.arange(nt) * dt
aGx = np.ones(nt-2) * gx; aGy = np.ones(nt-2) * gy
aDx = np.zeros(nt-2); aDy = np.zeros(nt-2)
vx = np.zeros(nt-1); vy = np.zeros(nt-1)
vx[0]=vel*np.cos(theta); vy[0]=vel*np.sin(theta)
```

As we have shown in the past chapter, we would like to completely vectorize every operation since they can be executed orders of magnitude faster than using loops. I will, therefore, vectorize all the loops in this section, with the exception of the time integration. Because the drag is proportional to the square of the speed, this cannot be vectorized if not solving analytically the associated differential equation, or by using an iterative approach starting from an initial guess of the entire time dependent solution. We will see in the next chapter how to use an iterative approach for nonlinear problems can be the best method for very large setups. Once the time evolution of the velocity is found, every other operation can be vectorized, although it requires 3-4 vector calculations.

We can now solve the evolutionary equations in time. Check that this solver works for every value of the speed, except zero, because it would trigger a division by zero. If the initial velocity is not vertical and upward, this is however impossible:

```
for it in np.arange(nt-2):
    vMag2=vx[it]**2+vy[it]**2
    accDrag=0.5*Const*densityAir*Area*vMag2/mass
    aDx[it]=-accDrag*vx[it]/vMag2**0.5
    aDy[it]=-accDrag*vy[it]/vMag2**0.5
```

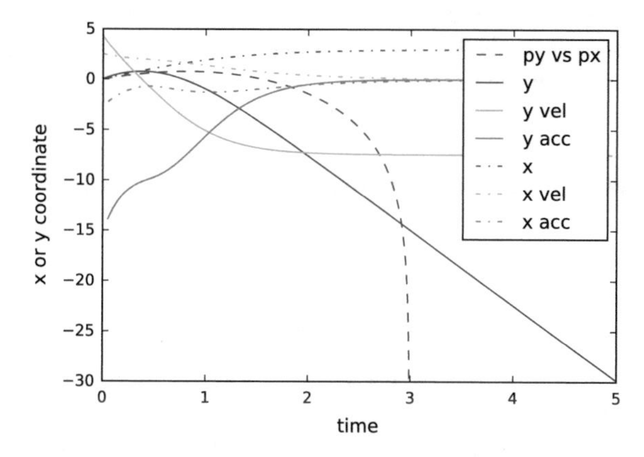

Fig. 5.2 Time evolution of x and y position, x and y velocity and x and y and acceleration versus time of a drop launched at an angle $\pi/6$ at $5\,\mathrm{m/s}$. The acceleration converges toward zero in about three seconds, which is the time necessary to the drop to reach steady state. The flex is here less pronounced because the speed never reaches the value of zero

```
    vx[it+1] = vx[it]+(aGx[it]+aDx[it])*dt
    vy[it+1] = vy[it]+(aGy[it]+aDy[it])*dt
```

Now that the velocity has been calculated, the $x-y$ trajectory can be found using only vectorized operations. In such a simple case, to vectorize the entire problem is an overkill, but for high-resolution problem this optimization is essential. To formulate the solution in a vectorized form is also compact and more readable, almost like a set of equations.

Because the next position depends on the past one, the time integration can done very efficiently by using the `cumsum` function of NumPy. The starting position of the trajectory (0, 0) is added using the `append` function:

```
pxInc = 0.5*(vx[:-1]+vx[1:])*dt # mid values for velocities
pyInc = 0.5*(vy[:-1]+vy[1:])*dt
px=np.append(([0.]),pxInc.cumsum()) #starts from 0.
py=np.append(([0.]),pyInc.cumsum())
```

The trajectories, i.e., the x and y positions, and the velocities and accelerations are plotted in Fig. 5.2, using MatPlotLib. One observes how the terminal speed is reached in 2 s from the moment in which the drop has reached the highest point in its trajectory, proving how the falling speed of rain drops is very modest.

5.1.3 Potential, Dissipated, Kinetic, Mechanical Energies for the Droplet

We can now calculate the energies associated to the system above and verify that the program correctly predicts that the sum of the mechanical energy and dissipated energies do not vary during the evolution of the trajectory. This kind of *energy check* is in this case relatively straightforward, but it is less immediate when dealing with continuum mechanics.

The kinetic energy of a moving rigid object is $K_E = \frac{1}{2}m\mathbf{v}^2$ and can be immediately calculated as

```
KE = 0.5 * mass *(vx**2+vy**2)
```

The Gravitational Potential Energy $P_E = -mgy$ is the source of the energy for the entire system. When the objects falls either is the potential energy transformed into kinetic energy, by the acceleration of the drop, or it is transformed into heat, through the friction with the air, that we condense in the concept of *Drag*:

```
PE = mass*(-gy)*py
```

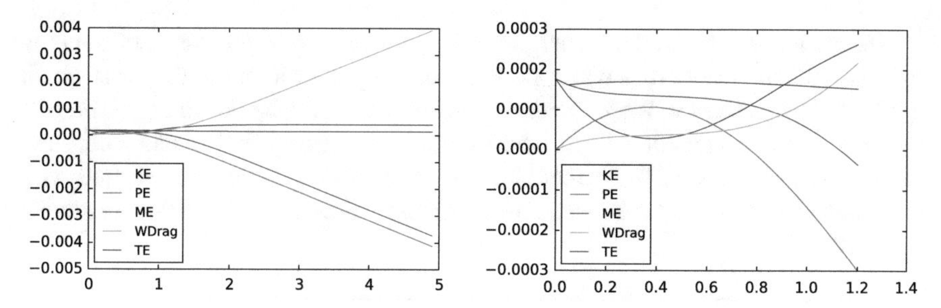

Fig. 5.3 *Left* Time evolution of Kinetic, Potential, and Mechanical Energies, compared with the dissipated Work (by air friction) and the verification that the Total Energy remains constant during the entire simulation. A zoomed view of the *left* part, where the energy exchanges happen is shown on the *right* plot

If the system is non-dissipative, i.e., for the droplet if the air exerts a negligible drag, the sum of the Kinetic energy and of the Potential Energy, called in general Mechanical Energy $M_E = K_E + K_E$, would be constant. Clearly in our case this law does not hold, as we can calculate here and plot it in Fig. 5.3.

```
ME = KE+PE
```

The Mechanical energy that is dissipated during the flight of the drop is instead converted into heat through friction. This energy can be calculated directly from the Drag Force and the kinematic variables, as it is the Work, the integral of the Power $P = \mathbf{D} \cdot \mathbf{v}$ in time: $W = \int \mathbf{D} \cdot \mathbf{v} dt = \int \mathbf{D} \cdot d\mathbf{x}$ where it is used the fact that $\mathbf{v} = d\mathbf{x}/dt$. The Drag Force itself can be immediately extrapolated from the kinematic solution, as $\mathbf{D} = m\mathbf{a}_D$. Notice how the vectorize version of the calculation requires the calculation of the increments first, to avoid looping:

```
pxInc = px[1:]-px[:-1]
pyInc = py[1:]-py[:-1]
WDragInc = -mass*(pxInc*aDx+pyInc*aDy)
WDrag = WDragInc.cumsum()
```

Finally, we calculate the total energy that has always to remain constant if all the energies are taken into account. It is the sum of the Mechanical Energy and the work done by the dissipative forces, the drag in our case: $T_E = M_E + W$. Notice how the Mechanical energy has to be calculated from the second element on. That is because the Drag is defined only when the velocity has been calculated, for the second element.

```
TotalEnergy = ME[1:] + WDrag
```

We can finally plot all these energies in Fig. 5.3 and verify that the kinetic energy increases at small velocities when the Work done by the friction of the air is small, while at steady state the Work increases linearly like the Potential Energy. The correctness of this calculation is verified by the Total Energy that remains constant during the dynamics. How correct is our numerical model can be measured with only one number, the ratio between the maximum variation in Total Energy and its mean value:

```
print((TotalEnergy.max()-TotalEnergy.min())/TotalEnergy.mean())
```

Which gives about 10^{-14} on my computer, which implies that the numerical scheme is very solid and reliable, even for a small number of intervals.

I did not obtain this value at the first attempt. I had to rewrite and refine the solver several times, until I obtained. It would be interesting here to go through all the errors that one could make while developing such a solver, but it would take us off from the main goal of this book, which is to develop geodynamic numerical tools. It is, however, essential to understand for the beginner who starts working on modeling that to obtain such a precision is essential (and sometimes not sufficient) to be certain of having developed a reliable code. To obtain it requires careful work and rethinking at the developed algorithm, often rewriting the code several times. One can see here that to be a able to write a compact and fast solver in Python is essential. We can anticipate that verifications of this sort (energy, entropy) on the more sophisticated continuum mechanics system that we will consider in the next chapters will be essential to check the validity of our simulations.

To improve the above algorithm, I had to invest time to reflect at how intermediate values had to be calculated (for velocities and acceleration). In particular, since Potential Energy depends on position and Kinetic Energy on velocity, typically they are displaced of half time. The work, if calculated using accelerations and positions is displaced of one interval from the mechanical energy, as shown in the calculation of the Total Energy. We will see in the next chapter on Lagrangian Dynamics, how to take *half time step* in consideration is essential in general to understand how bodies move and deform. Problem 5.2 offers also more ideas on how to improve the integration scheme.

5.2 Monte Carlo Simulation of the Pyroclastic flow During the 1944 Mt Vesuvio Volcanic Eruption

My father was born in 1941, just two years before the last eruption of Mt Vesuvio in March 1944, in the mid of the Second World War (Fig. 5.4). Although my father was just three- years old at the time of the eruption, he use to repeat the story of how he vividly remembered large stones falling from the Volcano to his village, about 5 km away from the crater. The stones were later collected in a little *hill* at the center of the village and children used it as their preferred playground for many of the following years.

Fig. 5.4 Image from the U.S. National Archives of the eruption of the Mt Vesuvio. One notices that the explosion was strongly asymmetric. This fits well with the Monte Carlo model below showing that the large ones are expected to reach several km of distance. Several villages on that side of the volcano were completely destroyed

By the stories of my father these stones had a diameter of tens of centimeters. I often thought that it was their size that allowed them to reach such a great distance, as well as the height of the Vesuvio (about 1280 m). The trajectory of the lapils generated by of a volcano, if no asymmetries are taken into account, can be simulated with the 2D approach that we just developed.

An exercise like this one, in which one explicitly simulates a phenomena many times and studies its statistical properties is called *Monte Carlo*. The idea is to start with some initial distribution of the unknown (or just variable) properties, and run a sufficiently large number of simulations so that a clear conclusion can be inferred. In general Monte Carlo numerical experiments are a broad class of algorithms based on repeated random sampling. The main idea is that when a problem is very complex and possibly unpredictable, by running a large set of random cases that cover the entire range of possibilities one can *visualize* and *calculate* the desired answer.

For example, to verify the story of my father I first created a routine written like the code developed above, and called *SingleLaunch(py0, radius, density, theta, vel)* where the variables indicate initial vertical position (the height of the crater), the radius of the lapil of which we want to calculate the trajectory, its density, *theta* the angle at which it leaves the volcano and *vel*, its speed. The code that calculates about 5000 trajectories looks like this:

```
launches = 5000
landing = np.zeros(launches)
radii = np.zeros(launches)
mass = np.zeros(launches)
for thisTrajectory in np.arange(launches):
    radius=random.uniform(0.05, 0.5)
    density=2500.
```

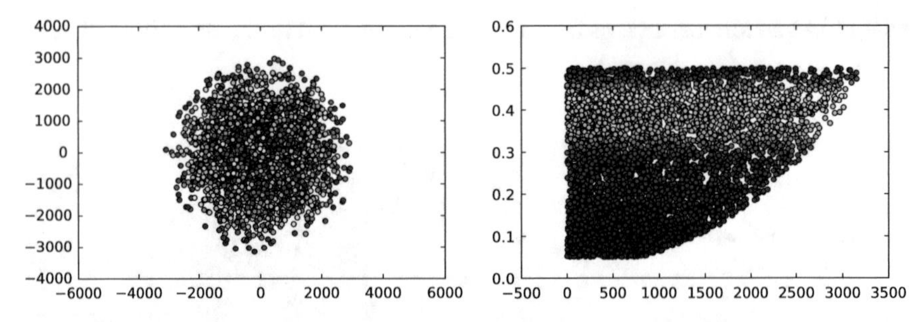

Fig. 5.5 *Left* Distribution around the crater of 5000 simulated lapils. The color represent the *radii* of the lapils, going from *blue* (low radius) up to *red* (high radius). The axis visualize distances in meters. Right: Plot of the Radii (in meters) versus Distance (in meters) reached by the lapils. Here every dot is one launch. The colors here represent the mass of the lapil

```
theta=random.uniform(0.,np.pi/2)
vel=random.uniform(10.,200.)
py0 = 1300.

(px,py)=SingleLaunch(py0,radius, density, theta, vel)

landing[thisTrajectory]=px[py<0][0]
radii[thisTrajectory]=radius
mass[thisTrajectory]=4.0/3.0*np.pi*radius**3*density
```

Notice that I can immediately detect the landing distance from the volcano with the instruction $px[py < 0][0]$, which is the horizontal position px at the first element of the array in which $py < 0$. The landing position, radii of the lapils, and angle and velocities in which they leave the volcano can now be plotted and the results is in figure Fig. 5.5.

By testing this case, and playing with the parameters, one can reach interesting conclusions. The first is that only large and heavy lapils can reach great distance, because they beat the resistance of the air. The second is that the ones that reach the largest distances arrive at around 3000 m of distance. This is not totally at odd with the memory of my father, but it seems to require some extra energy to reach 5 km distance. Clearly for scientific purposes one should look at historical reliable data, not child memories, but this is only a good reason to learn a new technique. One can observe that by adding wind in the forcing terms of the routine, one can probably obtain important differences in the result, in particular for the small grains, like dust, that will be heavily transported by the wind. I leave this case as an exercise for the motivated student.

5.3 Precession of a Gyroscope

As discovered by Kepler, the planets orbit on ellipses with the Sun at one focus, all revolving in the same direction. At the same time they spin around their axis. Newtonian dynamics teaches us that every object that is rotating around its axis in the space is also precessing at a much slower speed that depends on the 3 principal moments of inertia. We can use the techniques just learned to model 1D Newtonian Dynamics to model how the rotation of a planet generates precession.

For rotating objects, the linear momentum equation (5.2) have an angular analogue where *force* is replaced by *torque* and linear momentum by angular momentum:

$$\mathbf{L} = I\omega \qquad (5.4)$$
$$\mathbf{T} = \frac{d\mathbf{L}}{dt} = I\frac{d\omega}{dt}$$

where $\mathbf{T}$ is the Torque vector, equal to $\mathbf{T} = \mathbf{r} \times \mathbf{f}$ and $\mathbf{L}$ is the angular momentum vector. In analogy to linear momentum, the angular momentum is the tendency of an object to continue to rotate around its present axis and at its present rotating speed. Formally the angular velocity is the time derivative of the angular displacement $\omega = d\theta/dt$, taking the rotation axis that passes through the center of mass. The moment of Inertia depends instead from the shape of the object and it is proportional to its mass.

A peculiar characteristic of rotational dynamics is the precession. While objects moving in a linear trajectory can only accelerate or decelerate in the three directions of the space, a rotating object, if undergoing a force nonaligned with its axis, will display a dynamics characterized by two (or more) rotational dynamics, one around its original axis and the others around other axis. Normally these rotations are slower. Precession is one of them.

A gyroscope precesses (or processes) due to the torque applied by the force of gravity to the inclined axis of rotation. We will quantify this complex process by and visualizing in 3D using MatPlotLib. In order to do that let us use an extra module that defines a class of 3D arrows objects. Patches associated to the special arrow are here `http://matplotlib.org/api/patches_api.html`.

```
from numpy.linalg import norm
from matplotlib.patches import FancyArrowPatch
from mpl_toolkits.mplot3d import proj3d

class Arrow3D(FancyArrowPatch):
   def __init__(self, xs, ys, zs, *args, **kwargs):
      FancyArrowPatch.__init__(self, (0,0), (0,0), *args, **kwargs)
      self._verts3d = xs, ys, zs
   def draw(self, renderer):
      xs3d, ys3d, zs3d = self._verts3d
      xs, ys, zs = proj3d.proj_transform(xs3d, ys3d, zs3d, renderer.M)
```

```
        self.set_positions((xs[0],ys[0]),(xs[1],ys[1]))
        FancyArrowPatch.draw(self, renderer)
```

First, we need to define all the constants. The mass and moment of inertia of the gyroscope are given. `theta` is the initial angle θ of inclination of the gyroscope and `omega` the angular velocity ω. `L` and `r` represent the time evolution of 3D vectors, so they are 2D arrays.

```
m=1.0; I=1.0; theta=-np.pi/20
g=np.array([0.,0.,-10.0]); omega=10 #rad/s
nt=101; timeTot=2*np.pi; dt=timeTot/nt; time=np.arange(nt)*dt

L=np.zeros((nt,3),float)
r=np.zeros((nt,3),float)
```

We can further simplify the equations based on the fact that $\mathbf{L} = I\omega = I\omega\mathbf{r}$, where we assumed that the angular speed ω does not change in time (a real gyroscope will instead slow down due to frictional forces, but we neglect it as we want to apply this problem to the precession of the Earth). Because the ω does not change, only $\mathbf{r}$ will evolve in time.

Initially the axis of rotation is $\mathbf{r} = (0., \sin(\theta), \cos(\theta))$, where obviously $|r| = 1$. The gravity force can be written as $\mathbf{f} = m\mathbf{g}$.

```
F=m*g
r[0]=np.array([0.,np.sin(theta),np.cos(theta)])
L[0]=omega*I*r[0]
```

Let us now express the Angular Momentum as in Eq. (5.4) by reformulating the increment in time of the angular momentum as $d\mathbf{L} = \mathbf{T}dt$. Because $\mathbf{T} = \mathbf{r} \times \mathbf{f}$, we can just add this quantity at every time step and calculate the evolution of the angular momentum. In finite differences this becomes:

$$\Delta\mathbf{L} = \mathbf{T}\,\Delta t = (\mathbf{r} \times \mathbf{f})\,\Delta t \tag{5.5}$$

A first intuitive calculation of the evolution in time of the Angular Momentum can be done as a *forward* integration. If we do that the formulation is very simple:

```
for it in arange(1,nt):
    Tau=cross(r[it-1],F)
    L[it]=L[it-1]+Tau*dt
    r[it]=norm(r[it-1])*L[it]/norm(L[it])
```

The solution with this approximation is on the left of Fig. 5.6, and is not what we wanted because the axis of the rotation changes inclination. This happens because, like for the calculation of the trajectory, the evolution of the angular momentum requires the correction of a posteriori of the calculation of the torque, that in itself depends on the axis of rotation. The corrected formulation of the `for` loop therefore becomes:

```
for it in arange(1,nt):
    Tau=cross(r[it-1],F) # calculated with the old r
    L[it]=L[it-1]+Tau*dt
    r[it]=norm(r[it-1])*L[it]/norm(L[it])
    #correction a posteriori
    Tau2=cross(r[it],F) # calculated with the new r
    L[it]=L[it-1]+0.5*(Tau+Tau2)*dt
```

This numerical solution is plotted on the right of Fig. 5.6. In this case, we obtain an acceptable solution using one single iteration. For applications to real problems successive iterations can be implemented until full convergence is obtained.

The solution can be compared also with the analytical one (e.g. [14]), that arises from the estimation of the precession speed $\Omega = d\phi/dt$:

$$d\phi = \frac{dL}{L\sin\theta} = \frac{T\,dt}{L\sin\theta} = \frac{Mgr\sin\theta\,dt}{I\omega\sin\theta} = \frac{Mgr\,dt}{I\omega} \tag{5.6}$$

Therefore $\Omega = \frac{Mgr}{I\omega}$. With our (carefully crafted) parameters $\Omega = 1.0\,\text{rad/s}$, which implies that in a time interval of 2π s as in our case there will be exactly one precession.

The plots have been obtained using the above introduced arrow in the following way:

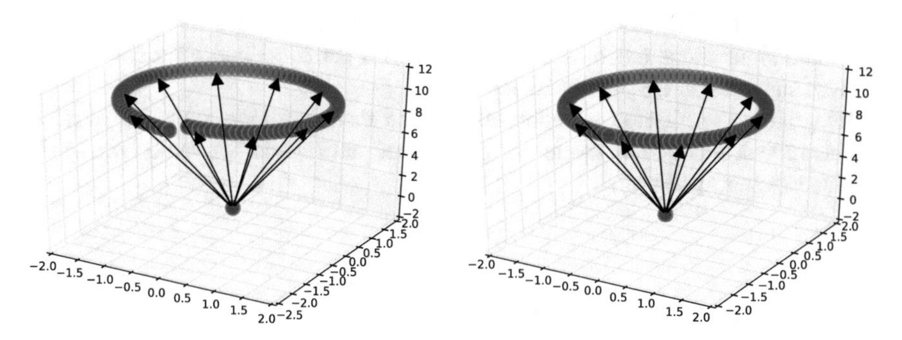

Fig. 5.6 Evolution of the gyroscope plotted as the position in time of the angular momentum $\mathbf{L}$. *Left* simulation using a simple *forward* approximation in which the increment $\Delta\mathbf{L} = (\mathbf{r} \times \mathbf{f})\,\Delta t$ is calculated using the past value of $\mathbf{r}$. *Right* $\mathbf{r} \times \mathbf{f}$ is calculated here with the $\mathbf{r}$ calculated on the simulation on the *left*, i.e., through one iteration. The simulation result looks already very good, i.e., consistent with the analytical solution. A better quantification of the error done can obtained by quantifying the distance between one iteration of $\mathbf{r}$ and the next one

```
fig = figure()
ax = fig.gca(projection='3d')
ax.scatter(L[:,0],L[:,1],L[:,2],color="g",s=100)
ax.scatter(0,0,0,color="g",s=100)

for it in arange(1,nt,10):  #plot every 10 steps
   a = Arrow3D([0,L[it,0]],[0,L[it,1]],[0,L[it,2]],
      mutation_scale=20, lw=1, arrowstyle="-|>", color="k")
ax.add_artist(a)
```

For the Earth, the precession is extremely slow and controlled by the difference between the Principal Momentum of Inertia that goes from one pole to the other, and one of the principal ones associated to one of the axis that cross the Earth at the equator. The velocity of precession proportional to this difference, which makes the speed of precession of the order of 26 thousands years.

Summary

- Linear Momentum and Conservation of Energy equations allow to solve any problem in Newtonian Mechanics, however only few can be tackled analytically.
- Calculating the trajectory of a body traveling in air, combining gravitational and air drag is a sufficiently complex example for testing integration techniques for solving the Newton equation.
- Calculating Energies (Dissipated, Kinematic, Mechanical) allows to monitor the finest errors that might be present and remain undetected in our simulations.
- Monte Carlo Methods are a very general suit of techniques that allow to use simple simulations (such as the trajectory of a particles) and infer conclusions from statistics on the possible outcomes.
- Integrating the Newton equation allows also modeling more complex physical phenomena such as the motion of gyroscope, that follows the same laws of the precession of the Earth.

Problems

5.1 The technique illustrated in Sect. 5.1.2 can be extended in many directions. A simple one is to investigate the time necessary to reach the ground by varying the angle of launch of the droplet and the viscosity of the atmosphere. If there was no air, and therefore the atmosphere viscosity was null, the speed at the ground of the droplet would be independent on the launch angle, due to conservation of mechanical energy. One can therefore calculate analytically the time required to reach the ground in this case, which represents a lower bound for the travel time. For a viscosity as large as the one of the Earth's atmosphere the velocity reaches very quickly a steady state value, therefore this is an upper bound (if we exclude atmospheres more viscous than Earth, such as on Venus).

The exercise is to transform the above technique into a *Python function* and use it to study to find the parametric dependence of the arrival time for a variety of viscosities and launch angles. Plot the result in 2D, where the angle is at the x-axis and the viscosity at the y-axis.

5.2 Some quantities in the simulations in Sect. 5.1.3 are calculated at time $t, t+1, t+2$ and so, while others at time $t+1/2, t+1+1/2, t+2+1/2$ and so on. Find out which ones are calculated at one time and which ones at the other time, and try to change where the average is calculated and notice how this perturbs the solution. In particular note how the total energy conservation is heavily influenced by the smallest error in the integration scheme. A quick way to study this problem is to create a function that solves the problem and calculates
`(TotalEnergy.max()-TotalEnergy.min())/TotalEnergy.mean()`
for every parameter choice.

5.3 This is a more difficult and advanced problem. The earliest stages of planetary evolution are believed to be characterized by a melted magma ocean through which iron rich diapirs fall toward its interiors and form the planetary core. Many recent researches aim at understanding how the sinking of these diapirs took place, in particular how large these diapirs were and how the interaction with the surrounding magma shaped, or broke up. The exercise is the following: assuming a drag Coefficient of 0.1, 1, 10 and 100 and using the same drag coefficients assumed in the Sect. 5.1.3, calculate the dynamics before reaching equilibrium velocity for diapirs of size of 10^{-3}, 10^{-2}, 10^{-1}, 1, 10 and 100 m. The other parameters that you will need are: (1) magma ocean viscosity: 1 Pas, (2) magma density 3800 kg/m^3, and (3) iron density 7800 kg/m^3. Assume that gravity was like today, about 10 m/s. You can compare your solution with the thickness of our mantle (about 3000 km) and make your model more physical from scientific literature (e.g., [82], Sect. 3).

5.4 The implementation of the Monte Carlo Simulation of a volcanic Eruption (Sect. 5.2) can be easily extended in 3D. By doing that, you can create a simulator and calculate the trajectory of the lapils ejected by a volcano during an explosive eruption. The goal of this exercise is to build a *Monte Carlo* simulator that starting from an initial probabilistic distribution of the directions and velocity of lapils coming from the volcano, builds a probability map of the location where the lapils will reach the ground. While the above model is simple and effective in 2D, a 3D model would allow to test the effect of the wind, non spherical shape, and others possibly important variables. What do you need to reach the distance of 5 km?

Chapter 6
Physics of Stokes Flow

"The only thing that you absolutely have to know, is the location of the library." — Albert Einstein

Abstract Some key concepts on the physics of Stokes (i.e., low Reynolds number) flow are introduced. Stokes flow presents many paradoxes, in particular in its 2D formulation. For example, in 2D the drag on a moving particle will always strongly depend on the boundary conditions, even if very faraway. Ultimately this emerges by the fact that Stokes flow is the end member case of Navier Stokes for infinitively fast vorticity diffusion. Fundamental solutions of Stokes flow (Stokeslet, Stresslet, Rotlets) allow writing the solution of a sphere and any other body as a combination of them. Einstein viscosity can be shown to result from the average for the contribution for a large number of spheres. For high packing the viscosity increases, until jamming.

Until now we have dealt only with moving solid bodies, either in translational or rotational mode. But what happens when bodies can deform? In this case we have to rewrite the equations of motions for fully deformable bodies. The two equations that will allow to solve most fluid-mechanics problems are the continuity equations and the *continuum version* of the momentum equation. Let's derive them for a cube material.

6.1 Momentum and Continuity Equations

The first concept that one has to relate with when working in continuum mechanics is the one of *stress*. The *stress tensor* is a 3×3 array in 3D, or a 2×2 one in 2D, that defines the internal forces inside deformable object. The stress tensor in general has the form:

G. Morra, *Pythonic Geodynamics*, Lecture Notes in Earth System Sciences,
DOI 10.1007/978-3-319-55682-6_6

$$\begin{bmatrix} \sigma_{xx} & \sigma_{xy} & \sigma_{xz} \\ \sigma_{yx} & \sigma_{yy} & \sigma_{yz} \\ \sigma_{zx} & \sigma_{zy} & \sigma_{zz} \end{bmatrix} \tag{6.1}$$

and its meaning is related to the one of the traction **t**. Taken an asymptotically small portion of material in space, the traction is the inner product of the stress tensor for the normal to the surface of the small sample:

$$t_i = \sigma_{ij} n_j \tag{6.2}$$

its dimensions are the one of pressure (N/m^2). Its physical meaning is clear when imaging the material portion to have a cubic shape (e.g. Fig. 6.1). In this case the product of the traction for the face area represents the force on that face in the direction of the traction. This means that the traction perpendicular to the face is the pressure on that face, while the 2 forces parallel to that face are shear forces.

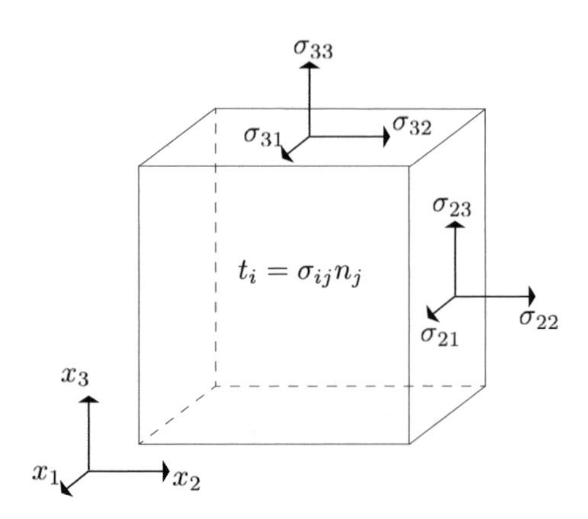

Fig. 6.1 Sketch of the meaning of the stress tensor of a small cube immersed in a deformable body. The traction vector is the inner (dot) product of the stress for the normal to the face, and gives the three components shown in the figure

The Newton's law, $\mathbf{f} = m\mathbf{a}$, or more in general $\mathbf{f} = \frac{d\mathbf{p}}{dt}$ can be expressed for this little cube by taking into account that the force can be either a *surface force* applied from the surrounding matter, through the stress multiplied to the surface, or as a *body force*, like gravity, or the centripetal force. Combining all these information together and considering that the linear momentum per unit of matter is $\mathbf{p} = \rho\mathbf{v}$, one obtains an expression where the force is given by the sum of a surface and a volume integral:

$$\int_S \sigma_{ij} n_j dS + \int_V \rho b_i dV = \int_V \rho \frac{dv_i}{dt} dV \tag{6.3}$$

where Einstein summation was used for the repeated index j. In order to eliminate the integral we need to transform the surface integral into a volume one. This can be done by exploiting the divergence theorem that states that the surface integral of the outward flux of a vector field through a closed surface is equal to the volume integral of the divergence over the region inside the surface. Since $\sigma_{ij} n_j$ is the outward flux of the field σ_{ij}, then one can rewrite the momentum equation as:

$$\int_V \frac{\partial \sigma_{ij}}{\partial x_j} dV + \int_V \rho b_i dV = \int_V \rho \frac{dv_i}{dt} dV \tag{6.4}$$

And since the integral holds for every volume of space, one can write the general momentum equation in continuum mechanics as:

$$\frac{\partial \sigma_{ij}}{\partial x_j} + \rho b_i = \rho \frac{dv_i}{dt} \tag{6.5}$$

which becomes, for the reader who is familiar with the tensor notation, the more compact expression:

$$\nabla \cdot \sigma + \rho \mathbf{b} = \rho \frac{d\mathbf{v}}{dt} \tag{6.6}$$

This is the general expression of the *momentum equation* for an external body force $\rho\mathbf{b}$.

The stress tensor has several characteristics. A fundamental one is that it is symmetrical, i.e. $\sigma_{ij} = \sigma_{ji}$, which is a consequence of the conservation of the angular momentum. In many textbooks of continuum mechanics, this stress is called *Couchy stress*, and several other types of stress are defined (e.g. Kirchhoff stress, Nominal Stress, Material Stress), however in this book stress will always only mean *Couchy stress*.

The momentum equation is always accompanied with a sister conservation equation, in general the continuity one. This can be found observing that the equilibrium between mass fluxes and density in a given set (eulerian) volume is in the integral form:

$$\int_V \frac{\partial \rho}{\partial t} dV + \int_S \rho v_i n_i dS = 0 \tag{6.7}$$

where $\rho v_i n_i$ is the flux through the closed surface S. This can be also transformed into a volume integral, as in (6.4), to obtain:

$$\int_V \frac{\partial \rho}{\partial t} dV + \int_V \frac{\partial \rho v_i}{\partial x_i} dV = 0 \tag{6.8}$$

and eliminating the integral one obtain the expression:

$$\frac{\partial \rho}{\partial t} + \frac{\partial \rho v_i}{\partial x_i} = 0 \tag{6.9}$$

This expression uses the partial derivative in time. A more common way to express it is by using the total time derivative of $\rho(\mathbf{x}, t)$, which is:

$$\frac{d\rho}{dt} = \frac{\partial \rho}{\partial t} + \frac{\partial \rho}{\partial x_i}\frac{\partial x_i}{\partial t} = \frac{\partial \rho}{\partial t} + v_i \frac{\partial \rho}{\partial x_i} \tag{6.10}$$

By splitting the derivative of the product in the second term of (6.9), one rewrite the equation (6.9) as:

$$\frac{d\rho}{dt} + \rho \frac{\partial v_i}{\partial x_i} = 0 \tag{6.11}$$

or, in the more compact tensorial form:

$$\frac{d\rho}{dt} + \rho \nabla \cdot \mathbf{v} = 0 \tag{6.12}$$

This is the general expression of the *continuity equation*. The most common approximation done in fluid-dynamics is to assume that the variation of density are small enough (incompressibility) that one can assume $\frac{d\rho}{dt} = 0$ and therefore write this fundamental equation simply as $\nabla \cdot \mathbf{v} = 0$.

6.1.1 Navier Stokes Equation

By combining the *momentum* and *continuity* equations with a constitutive expression for the stress tensor, it is possible to obtain the general Navier-Stokes equations that describe the motion of every viscous fluid. This can be done introducing a the concept of strain rate. The physics meaning of strain rate will be better clarified in Sect. 6.3. Here we just define it in function of the velocity field $\mathbf{v}$, whose components are v_i:

$$\epsilon_{ij} = \frac{1}{2}\left(\frac{\partial v_i}{\partial x_j} + \frac{\partial v_j}{\partial x_i}\right) \tag{6.13}$$

Viscous fluids are defined by direct relationship between stress and strain rate. The simplest one is linear:

$$\sigma_{ij} = 2\mu\epsilon_{ij} - \delta_{ij}p \tag{6.14}$$

where δ_{ij} is one when the indexes are the same, and zero when different. In this expression the shear and bulk components have been split. The viscosity μ therefore expresses only the resistance to shear motion, not to bulk deformation.

Traditionally Navier Stokes equations are found for an incompressible fluid, therefore we can assume $\nabla \cdot \mathbf{v} = \partial_i v_i = 0$. By inserting (6.14) into (6.5) and using the definition of strain rate in (6.13), one can rewrite the divergence of the stress tensor in (6.5) as:

$$\frac{\partial \sigma_{ij}}{\partial x_j} = \frac{\partial\left(2\mu\epsilon_{ij}\right)}{\partial x_j} - \frac{\partial\left(\delta_{ij}p\right)}{\partial x_j} = \mu\frac{\partial}{\partial x_j}\left(\frac{\partial v_i}{\partial x_j} + \frac{\partial v_j}{\partial x_i}\right) - \frac{\partial p}{\partial x_i} \tag{6.15}$$

where I assumed that the viscosity of the fluid does not depend on $\mathbf{x}$. This is in general not correct for geomaterials, for which the effective resistance to shear is great. However I will consider this case when building a non-linear solver for the Stokes. One advantage of this formulation is that partial derivatives in x_i and x_j can be swapped, which allows using the continuity equation to eliminate one of the derivatives of the velocity term:

$$\frac{\partial}{\partial x_j}\left(\frac{\partial v_i}{\partial x_j} + \frac{\partial v_j}{\partial x_i}\right) = \frac{\partial}{\partial x_j}\frac{\partial v_i}{\partial x_j} + \frac{\partial}{\partial x_i}\frac{\partial v_j}{\partial x_j} = \frac{\partial}{\partial x_j}\frac{\partial v_i}{\partial x_j} = \frac{\partial^2 v_i}{\partial x_j^2} = \nabla^2 v_i \tag{6.16}$$

We can therefore now write the classical expression for the *Navier-Stokes* equation for every v_i component:

$$\rho\frac{dv_i}{dt} = \mu\frac{\partial^2 v_i}{\partial x_j^2} - \frac{\partial p}{\partial x_i} + \rho b_i \tag{6.17}$$

or in a more elegant, perfectly equivalent, vectorial form:

$$\rho\frac{d\mathbf{v}}{dt} = \mu\nabla^2\mathbf{v} - \nabla p + \rho\mathbf{b} \tag{6.18}$$

The left term can be written in many ways by developing the total derivatives in partial derivatives:

$$\frac{dv_i}{dt} = \frac{\partial v_i}{\partial t} + \frac{\partial v_i}{\partial x_j}\frac{\partial x_j}{\partial t} = \frac{\partial v_i}{\partial t} + \frac{\partial v_i}{\partial x_j}v_j = \frac{\partial v_i}{\partial t} + (\mathbf{v}\cdot\nabla)\,v_i \tag{6.19}$$

Which in vectorial form transforms (6.18) in

$$\rho \frac{\partial \mathbf{v}}{\partial t} + \rho (\mathbf{v} \cdot \nabla) \mathbf{v} = \mu \nabla^2 \mathbf{v} - \nabla p + \rho \mathbf{b} \tag{6.20}$$

In Stokes Flow the two terms on the left are assumed to be negligibly small. The magnitude of these terms is proportional to the Stokes St number and Reynolds Re number, respectively. In fact

$$Re = \frac{Va\rho}{\mu} \approx \frac{\rho |(\mathbf{v} \cdot \nabla) \mathbf{v}|}{\mu |\nabla^2 \mathbf{v}|} \tag{6.21}$$

where V is the velocity of the flow and a the size of involved. Similarly

$$St = \frac{a^2 \rho}{T\mu} \approx \frac{\rho |\partial \mathbf{v}/\partial t|}{\mu |\nabla^2 \mathbf{v}|} \tag{6.22}$$

where T is the timescale of the process. For all the models that will consider in this book we will assume that $Re \ll 1$ and that $St \ll 1$, and therefore will solve the Stokes equation:

$$\mu \nabla^2 \mathbf{v} - \nabla p + \rho \mathbf{b} = 0 \tag{6.23}$$

Which appears a very simple equation, but in reality hides many paradoxes. Furthermore, if the approximation of homogeneous viscosity does not hold anymore, it assumes a much more complex form, that we will investigate numerically in Chap. 10.

6.2 Stokes Flow: Simple but Not Obvious

Stokes flow is described by simple and linear equations, however observation tells us that in many cases its behavior is unexpected. Let us look at some remarkable examples to develop an intuition of some of its hidden aspects.

6.2.1 Stokes' Paradox

One of the best known analytical solutions in physics and geosciences is the drag on a sphere due to a highly viscous fluid. This force, called Stokes' Law, is $\mathbf{F} = -6\mu\pi a\mathbf{v}$ where μ is the viscosity of the fluid, a the radius of the sphere and $\mathbf{v}$ the velocity of the moving particle. The minus indicates that the drag opposes the motion of the particles. This is the calculation of the drag on sphere, as we introduced in Chap. 5, but in the limit of vanishing Reynolds number.

A natural question that one could ask is: why did Stokes found a 3D analytical solution of the problem (in 1851!) but not a 2D solution? If you did, you would not be the only one. About thirty years later Whitehead, in 1889, documented his long attempts to find the solution for an infinitively long cylinder, which would be the 2D case, without success. Such a surprising fact is now called sometimes *Stokes' paradox*, other times Whitehead paradox. Whitehead in practice realized that the same mathematical proof that works in 3D does not converge in 2D, i.e., that the force on an infinitely long cylinder immersed in a highly fluid moving transversally to the axis of the cylinder is theoretically infinite.

The paradox was solved only after 30 more years by Oseen, a physics professor from Uppsala, known for his elasticity theory of liquid crystals. Oseen showed that the linear momentum of the flow moving very far away from the infinite cylinder is never negligible, and that therefore the approximation used to write the flow law of Stokes Flow is not correct for an unbounded 2D flow. In practice, it is impossible to find a $2D$ Stokes' Drag because this force does not exist in an unbounded domain.

On the other side if the cylinder is not in unbounded domain, but in a walled one, then there is always a solution, which depends indeed in the boundary. Strikingly the dependence of the drag on the smallest cylinder does never vanish even if its thickness is many orders of magnitude less than the wall. When later we will have to benchmark our continuum mechanics codes that will be all initially developed in 2D, such as in Chap. 10, it will be very important to take this paradox into account.

The analytical solution of an infinitely long cylinder drifting perpendicularly to its axis, inside an infinitely long cylindrical pipe (sketch in Fig. 6.2). For this system the drag of the flow to the inner cylinder is

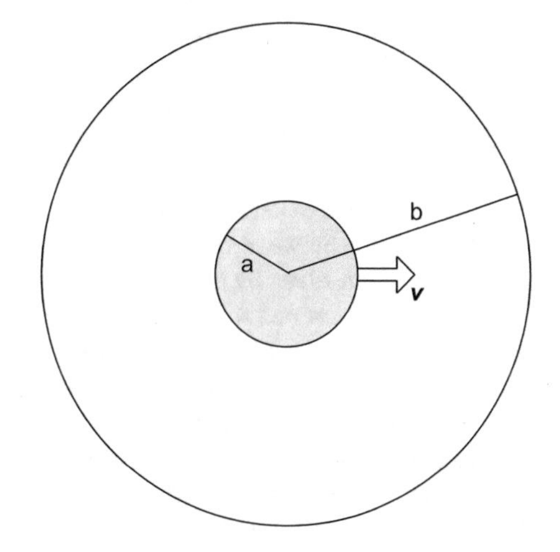

Fig. 6.2 The *inner* rod of radius a is immersed in a fluid that separates it from the cylinder of radius b, while moving perpendicularly to its axis. Even for tiny a/b ratios the drag force applied to the rod will always depend on the distance from the cylinder

$$\mathbf{F} = -\frac{4\mu\pi\mathbf{v}}{log\left(\frac{b}{a} - 1\right)} \tag{6.24}$$

where b is the radius of the cylinder and a is the radius of the rod. This equation tells us that when b goes to infinite, i.e., the rod is immersed in an unbounded fluid, the drag goes to infinite as well, which is the Stokes' paradox, and means that the Stokes flow approximation is not valid for such a system. This solution is shown in Fig. 6.2.

This formula shows that it does not matter whether the rod is orders of magnitude smaller than the cylinder, the drag on the rod will always depend on the ratio between cylinder size and the rod ratio. This implies that when modeling a motion in 2D in a confined system, we are always calculating the dynamics of an elongated object in the flow, that will always feel the boundaries, and not the flow of a 3D object, whose dynamics will instead be independent from the presence of boundaries that are sufficiently distant. Finally, this implies that if we want to reliably model the collective motion of objects in a fluid, we necessarily need to model it in 3D, since the interaction between them due to reciprocal drag in 2D will be much greater.

6.2.2 Flow Reversibility

The dynamics of mantle flow, lithospheric flow, flow in magma chamber, and many other geodynamic systems are controlled by Low Reynolds number hydrodynamics. However, an even greater field of application of this branch of fluid dynamics is microhydrodynamics. Very small-sized organisms have a tiny inertia, and for this reason they cannot glide effortlessly through the water like fishes. The Reynolds number $Re = \rho vl/\mu$ in fact is almost zero either for extremely low velocities v, as for most geodynamic problems, or for extremely tiny sizes l as for microorganisms. In a Stokes and Newtonian fluid, any movement based on a regular undulating movement like flapping fins will not bring any motion. This is due to the reversibility of the Newtonian Flow.

The way in which a microorganism swim is to change shape during motion in a nonreversible way. One trick is to use a helix. Elongated animals can in fact take a twisted shape and rotate, or spherically shaped organism can have a tail shaped as a rotating helix called flagellum (sketch in Fig. 6.3). In this way, exploiting the differential drag between the one longitudinal and perpendicular to the flagellum, they propel themselves by twisting the tail.

These considerations show us how the reversability property of the Stokes flow is very important, since it imposes stringent constrains on dynamics that are possible. It is however important to emphasize that this property is exclusively present when Low Reynolds number dynamics is combined with linear homogeneous viscosity. In fact if the viscosity is strain rate dependent, as in most fluids in biology and, approximated at a large scale, for the creeping of the Earth's mantle, reversibility does not hold anymore. Still it is very important to understand this property because the macroscopic nonlinear behavior of these fluids is often due to presence of elastic and/or strongly anisotropic particles immersed in an otherwise newtonian fluid. For this reason to understand how this nonlineary arises is essential in geological as well as biological sciences.

6.2.3 Origin of the Paradoxes

There is a simple way to comprehend why Stokes flow presents so many paradoxes, which means that its formulation is flawed in many aspects. This can be understood by reformulating the Navier– Stokes equation in terms of the vorticity $\boldsymbol{\omega}$. If incompressibility ($\nabla \cdot \dot{\mathbf{x}} = 0$) is assumed, the Navier–Stokes equation is rewritten in terms of the vorticity $\boldsymbol{\omega}(\mathbf{x}) = \nabla \times \mathbf{u}$ as:

$$\partial_t \boldsymbol{\omega} + \mathbf{u}\nabla\omega = \frac{\mu}{\rho}\nabla^2\boldsymbol{\omega} \tag{6.25}$$

where the second term is the advection of vorticity. When solving this equation at non negligible Reynolds number, its integration requires

For low Reynolds number, we can neglect the advection term $\mathbf{u}\nabla\omega$, implying that the Stokes equation assumes the very simple form of a diffusion equation:

$$\partial_t \boldsymbol{\omega} = \mu\nabla^2\boldsymbol{\omega} \tag{6.26}$$

What appears clear here from this formulation is that if the viscosity is not too high, the vorticity will diffuse until a certain distance, which means that we can assume a time step related to the diffusion term. In other words this shows us that taking zero Reynolds number is equivalent to assuming instantaneous diffusion everywhere in the space. This limit assumption is at the origin of all the paradoxes. This also means that the Stokes equation is not an *unphysical* assumption, but it only means that our domain of calculation has to be limited to a region not too large, were instantaneous diffusion is an acceptable approximation.

For most problems in geodynamics the diffusion coefficient is order of 10^{20} Pas, which implies that vorticity instantaneously diffuses everywhere in the space. In the Earth every deformation propagates at the speed of sound for rock materials (36 Km/s), which implies that it takes O(1 h) for a seismic wave to cross our planet.

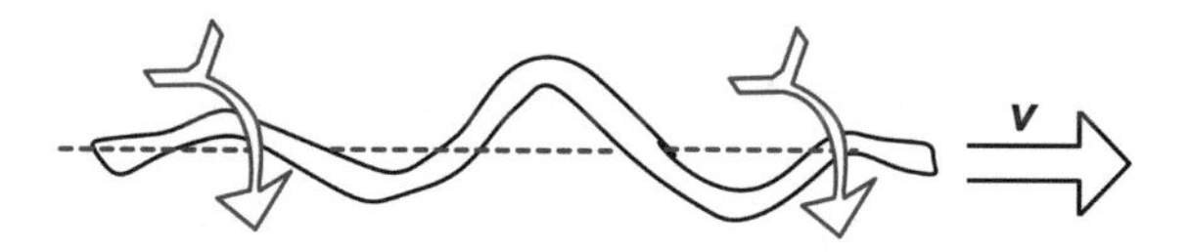

Fig. 6.3 A very low Reynolds number any repetitive reversed movement does not induce any movement due to the reversability of the Newtonian Stokes Flow. Microorganisms often move using flagellas that 'screws' through the liquid in which they are immersed

For every geodynamic calculation at 1 order of magnitude greater than that, i.e., at least of timestep of one day, Stokes Law is an acceptable geodynamic flow law.

6.3 Fundamental Solutions of Stokes Flow

Intuitively the best way to describe the motion of a fluid is to write it as the sum of a small number of *fundamental solutions* that capture the main behavior of the fluid. This is already routinely done, for example, using the spectral method, in which the spectrum indicates which fundamental sinusoidal modes mainly describe a certain function or signal.

In low Reynolds number fluid dynamics the fundamental solutions are the so-called Green Functions of the Stokes equations. They can be seen as the elements of a Taylor series of the velocity $\mathbf{u}(\mathbf{x})$ at a generic point $\mathbf{x}$ in the space:

$$\mathbf{u}(\mathbf{x}) = \mathbf{u}(\mathbf{x}_0) + \nabla\mathbf{u}(\mathbf{x}_0) \cdot (\mathbf{x} - \mathbf{x}_0) + \cdots \tag{6.27}$$

To describe the motion of a rigid sphere, the velocity of the sphere will be given by the above expression when $(\mathbf{x} - v x_0) = 0$. If we call the velocity of the sphere $\mathbf{U}$ we obtain $\mathbf{u}(\mathbf{x}_0) = \mathbf{U}$.

If we look now at the first-order term of the Taylor, we see that it is the product of the tensor $\nabla\mathbf{u}(\mathbf{x}_0)$, that is the gradiet of the velocity vector, with the relative position of the point in the space $\mathbf{x}$ respect to the position $\mathbf{x}_0$ of the sphere. We will see now that to split the tensor $\nabla\mathbf{u}(\mathbf{x}_0)$ in symmetric and antisymemtric components enlightens on the form of the most natural fundamental solutions of the Stokes flow from a physical point of view. Explicitly written in its component this tensor is

$$\nabla\mathbf{u}(\mathbf{x}_0) = \frac{\partial u_i}{\partial x_j} = \Omega_{ij} + \epsilon_{ij}$$

Where we called ϵ and Ω the symmetric and antisymmetric parts, respectively. Writing them Explicitly in their components they become

$$\Omega_{ij} = \frac{1}{2}\left[\frac{\partial u_i}{\partial x_j} - \frac{\partial u_j}{\partial x_i}\right] \qquad \epsilon_{ij} = \frac{1}{2}\left[\frac{\partial u_i}{\partial x_j} + \frac{\partial u_j}{\partial x_i}\right]$$

We recognize ϵ as being the *strain rate*, and Ω as being the *rotation rate*. Since the motion of the particles is described by $\mathbf{U}$, the sum of ϵ and Ω represents the *shear deformation* of the fluid surrounding the moving particle.

When its terms are explicitly written, an antisymmetric tensor for construction has the terms on the diagonal necessarily to be zero and the ones above the diagonal be the opposite of the ones below. For this reason, in 2D this tensor will have only one independent component and in 3D will have 3 independent components. Since three components can be comfortably packed into a vector, this has been done into the so called *rotation vector* $\boldsymbol{\omega}$ that is just half of the vorticity vector $\nabla \times \mathbf{u}$.

6.3.1 Rotlet

The flow induced by the rotation of a sphere is described by $\mathbf{u} = \omega \times \mathbf{x}$. If we want to calculate the disturbance flow $p(\mathbf{x})$ due to the rotation of the particle only, we need to solve the simplified Stokes flow equations:

$$\nabla \cdot \mathbf{u} = 0 \qquad \mu \nabla^2 \mathbf{u} = \nabla p \tag{6.28}$$

imposing the boundary conditions $\mathbf{u} = -\boldsymbol{\omega} \times \mathbf{x}$ at the boundary of the sphere, and $\mathbf{u} \to 0$ and $p \to 0$ for $| \mathbf{x} | \to \infty$. A detailed proof (see [12]) shows that the rotation induced pressure in an incompressible Netwonian fluid is zero and that the induced velocity can be described by

$$\mathbf{u}(\mathbf{x}) = \boldsymbol{\omega}^p \times \mathbf{x} \left(\frac{a}{r} \right)^3 \tag{6.29}$$

where a is again the radius of the sphere and $\boldsymbol{\omega}^p$ is the rotation rate of the particle. This solution is normally called `rotlet`, the fundamental solution of Stokes flow associated to rotation.

6.3.2 Stokeslet

Let us move now to the induced flow from a sphere that is translating in a flow that was otherwise quiescent. This is the basic problem that has to be solved in *sedimentation* and many other problems in geoscience. We again to solve the (6.28), this time with the boundary conditions $\mathbf{u} = \mathbf{U}^p$ at the boundary of the particle and again $\mathbf{u} \to 0$ and $p \to 0$ for $| \mathbf{x} | \to \infty$.

This solution can be found as before exploting the requirement for the pressure to be an harmonic function. Because the pressure must be also linear in the velocity of the particle $\mathbf{U}^p$, pressure must be proportional to both $\mathbf{U}^p$ and the second spherical solid harmonic $\mathbf{x}/r^3$:

$$p(\mathbf{x}) \propto \mathbf{U}^p \cdot \frac{\mathbf{x}}{r^3} \tag{6.30}$$

Summary

- The Stokes Paradox demonstrates that 2D and 3D Stokes flow are very different. In 2D, the solution will always depend on the Boundary Conditions. Not necessarily in 3D.
- Linear Stokes flow is always reversible. Nonlinear Stokes is however not so.
- Incompressible Navier–Stokes can be rewritten as a Diffusion equation, where vorticity diffuses in space. Stokes flow is the approximation of infinitively fast diffusing vorticity, which is the ultimate cause of the paradoxes.
- The fundamental solutions of Stokes Flow are Stokeslet, Stresslet, and Rotlet. Every motion can be represented by a combination of them.
- The solution of Stokes for a solid sphere and for an empty bubble can be written in function of the fundamental solution of Stokes flow.
- Einstein viscosity can be proven from the sum a large number of solid sphere in space far from each other. The viscosity becomes nonlinear by increasing the number of spheres, until jamming.

Part III
Lattice Methods

This part is composed of four chapters. In the first it is illustrated how to calculate derivatives on a Lattice using the finite-difference approximation. By using these techniques, the momentum and energy equations are extended to a continuum by defining and calculating strain, strain rate, and stress.

By solving the diffusion equation, it is shown the limitation of lattice-based numerical modeling, like numerical stability and numerical diffusion. The physical meaning of these limitations is also outlined.

Examples such as advection of particles, fields, and heat in a box are detailed together with a discussion of their validity. Finally, it is illustrated how to address nonhomogenous and nonlinear problems by the perturbation of linearized solutions.

Chapter 7
Lagrangian Transport

Abstract Lagrangian techniques are introduced to show how to calculate the internal motion of a convecting fluid in a box. Two types of flows are tested, (i) a rigid rotation of the space and (ii) a more complex thinning flow that stretches shapes. Emphasis is given to prevent the errors that might appear and quickly build up when calculating the trajectory of the Lagrangian particles. Three projections from the grid to the particles are illustrated, based on different weights. It is also illustrated how to implement these functions very efficiently in order to prevent lack of scalability and also use NumPy functions at their best. It is finally shown how to scale the projection up to a large number of particles and how to project back the field (Temperature, composition) from the particles to the mesh.

Until now we have only considered rigid objects that move in time. They could have been a falling water drop or the trajectory of a ship crossing a river. Most problems that are relevant in geodynamics, however, require calculating the internal deformation of solids and fluids. For example, the fracture of crustal material, the slow deformation of deep mantle, or the chaotic motion of iron in the outer core of the Earth. To understand these systems we need to define deformation quantitatively. Physics tell us that evolutionary laws are expressed as the time derivative of certain key quantities such as velocity, momentum and energy. To quantify deformation we need to combine derivation in space with derivation in time. For example, to quantify the internal deformation we use the so-called *Strain*, which is the change of displacement in a direction versus the space in the same (bulk) or another (shear) direction.

7.1 Strain and Strain Rate

To calculate Strain and Strain Rate, we need first to define the derivative in 2 and 3 dimensions of the vectors displacement and velocity. This is normally called gradient of a vector, indicated with the symbol ∇ and is expressed as

G. Morra, *Pythonic Geodynamics*, Lecture Notes in Earth System Sciences,
DOI 10.1007/978-3-319-55682-6_7

$$\frac{\partial \mathbf{s}}{\partial \mathbf{x}} = \begin{bmatrix} \frac{\partial s_x}{\partial x} & \frac{\partial s_x}{\partial y} \\ \frac{\partial s_y}{\partial x} & \frac{\partial s_y}{\partial y} \end{bmatrix} = \nabla \mathbf{s} \qquad \frac{\partial \mathbf{v}}{\partial \mathbf{x}} = \begin{bmatrix} \frac{\partial v_x}{\partial x} & \frac{\partial v_x}{\partial y} \\ \frac{\partial v_y}{\partial x} & \frac{\partial v_y}{\partial y} \end{bmatrix} = \nabla \mathbf{v} \tag{7.1}$$

The stress and strain rate are very similar. The only difference is that they are the symmetric component of them, they are indicated with the symbol ε (Strain) or $\dot{\varepsilon}$ (Strain Rate). Explicitly, in 2D, they are expressed by the average between the gradient expressed in (7.1) and its transpose. Strain therefore becomes:

$$\varepsilon = \frac{1}{2}\left(\nabla \mathbf{s} + \nabla \mathbf{s}^T\right) = \begin{bmatrix} \frac{\partial s_x}{\partial x} & \frac{1}{2}\left(\frac{\partial s_x}{\partial y} + \frac{\partial s_y}{\partial x}\right) \\ \frac{1}{2}\left(\frac{\partial s_x}{\partial y} + \frac{\partial s_y}{\partial x}\right) & \frac{\partial s_y}{\partial y} \end{bmatrix} \tag{7.2}$$

And strain rate:

$$\dot{\varepsilon} = \frac{1}{2}\left(\nabla \mathbf{v} + \nabla \mathbf{v}^T\right) = \begin{bmatrix} \frac{\partial v_x}{\partial x} & \frac{1}{2}\left(\frac{\partial v_x}{\partial y} + \frac{\partial v_y}{\partial x}\right) \\ \frac{1}{2}\left(\frac{\partial v_x}{\partial y} + \frac{\partial v_y}{\partial x}\right) & \frac{\partial v_y}{\partial y} \end{bmatrix} \tag{7.3}$$

The mathematical expressions in (7.2) and (7.3) are called *tensors*. Tensors have $n \times n$ components, where n is the number of dimensions. Despite their scary name tensors are very simple. When we derive a quantity in 2D, let us say velocity, we can derive v_x or v_y by either x or y, therefore there are 4 derivates, i.e., 2×2. Similarly in 3D we have 3 velocity components to derive by three axes, i.e., $3 \times 3 = 9$ components. Derivatives in n-dimensions, therefore, are objects that have n times components respect to the derived object. Space derivatives of the Strain Rate in 3D, for example, has $9 \times 3 = 27$ components. We will not deal with these objects here, but it is important to keep in mind that they exist and in computational geodynamics one may deal with them one day.

Before starting to solve continuum mechanics equations, we need to learn few important things. One is *Lagrangian transport*. Lagrangian means that the quantities transported (e.g. temperature or composition) are transported with the motion of the fluid, like tags on a particles. Eulerian, on the other side, means that the same quantities are defined on a fixed grid. The operation of projecting these information from tags that follow the fluid to/from the fixed grid is always diffusive. Diffusive here, it means that any *front* or any sharp boundary, will be *smoothed* by the projection process.

Diffusion has a particular important effect when we project the velocity itself from a cartesian grid to points in space that we want to transport. As we will see more in detail in Chap. 9, any mathematical expression involving second derivatives in space (as in every diffusion process, as well as in Stokes and Darcy flow) is expressed on a set of discrete points in space, its discretization itself produces extra terms that have the form of a wave equation. In other words every sharp boundary, when advected with a finite difference computational scheme, will become the *front* of a wave that propagates with its own properties, that depend on the sharpness of the front itself.

I will show some examples of this behavior in the rest of the book. In this chapter, I will instead focus on how advection can be performed by projecting the key quantities (velocity, temperature, composition) from the lattice to a set of moving particles. This method minimizes errors by reducing diffusion and avoids instabilities like the wave propagation.

7.2 Rigid Rotation

Rigid rotation is the motion of a system that revolves around an axis. It can involve only one body that rotates around an internal axis, which implies large variations of angular velocity between different parts of a body, or that rotates around a distant axis, in which case the rotational component is only a perturbation on a linear translation. Since every shape is preserved during motion, while at the same time nonlinearly crossing any background cartesian mesh, this form of transport is an excellent test for boundary transport algorithms.

Computer scientist have in fact envisaged many algorithms to deal with such a problem, all based on a variety of approximations. Before testing a set of methods, let us start from a simple setup, i.e., the transport of a set of points that regularly placed points defining a surface dividing two domains. The points will initially be placed in a circular shape. We assume that the fluid A is contained in the circle, while the fluid B is everywhere else. We will indicate from now on these *points* as *particles*, as they introduce us for the first time to the features of the general *Particle in Cell* numerical method.

A normalized expression of the velocity field associated to rotation is given by the simple expressions:

$$\begin{aligned} v_x &= -y \\ v_y &= x \end{aligned} \tag{7.4}$$

Let us create now a Python program that can transport the particles. After importing the necessary modules, we can create the 2D arrays X and Y, and calculate the velocity field in (7.4) on each node of the mesh. Let us call *nxp* and *nyp* the number of nodes (points) in the x and y direction, and $nxc = nxp - 1$ and $nyc = nyp - 1$ the respective number of intervals (cells). For simplicity, we assume that x and y both vary between -0.5 and 0.5:

```
import numpy as np
from numpy.linalg import norm
import matplotlib.pyplot as plt

nxp=101; nxc=nxp-1; nyp=101; nyc=nyp-1
xTot=1.0; dx=xTot/nxc; xMin=-xTot/2.0; xMax=xTot/2.0
yTot=1.0; dy=yTot/nyc; yMin=-yTot/2.0; yMax=yTot/2.0
```

```
# allocate the lattice arrays
X = np.arange(nxp, dtype=float)*dx+xMin   # x lattice coordinates
Y = np.arange(nyp, dtype=float)*dy+yMin   # y lattice coordinates

vx = np.zeros((nxp,nyp), float)   # x-velocity at nodes
vy = np.zeros((nxp,nyp), float)   # y-velocity at nodes

# initialize the velocity field
for iy in np.arange(0,nyp): vx[:,iy] = -Y[iy]
for ix in np.arange(0,nxp): vy[ix,:] = X[ix]
```

Where we have also allocated the memory for the velocities at the nodes. We do not mind about using here *for* loops because we use them only at the beginning to initialize the system. Still we want to avoid them for repeated tasks, and find a vectorized version of the code. The flow is represented by the blue lines in Fig. 7.1.

The idea of particle in cell method is proceed through four phases:

A. Interpolate fields from mesh (lattice) to particles
B. Move the particles
C. Project physical quantities from the particles to the mesh
D. Recalculate the fields on the mesh (lattice).

Since our goal is now to create an initial version of a particle in set code for Geodynamic problems, we will need to project the velocity field from the lattice (cartesian mesh) to the particles. Let us, therefore, allocate arrays for storing particle position and the projected particle velocity.

```
pN=100
px=np.zeros(pN, float)
py=np.zeros(pN, float)
vxt=np.zeros(pN, float)
vyt=np.zeros(pN, float)
```

For pedagogical purposes, it is useful to gradually increase the number of particles. We will initially test a very small set, just 100. In later applications we will fill the entire computational space and associated physical fields (such as composition and temperature) of particle. Since our particles will be in great number, it is essential that every operation on the particle is vectorized, to develop an already optimal code.

We will place the 100 particles to initially form the shape of a circle. This allows us to immediately monitor the evolution of its shape versus time. We will place the circle with the center at $1/4$ to the side of the domain center, with a radius equal to $1/10$ of the length of the domain side.

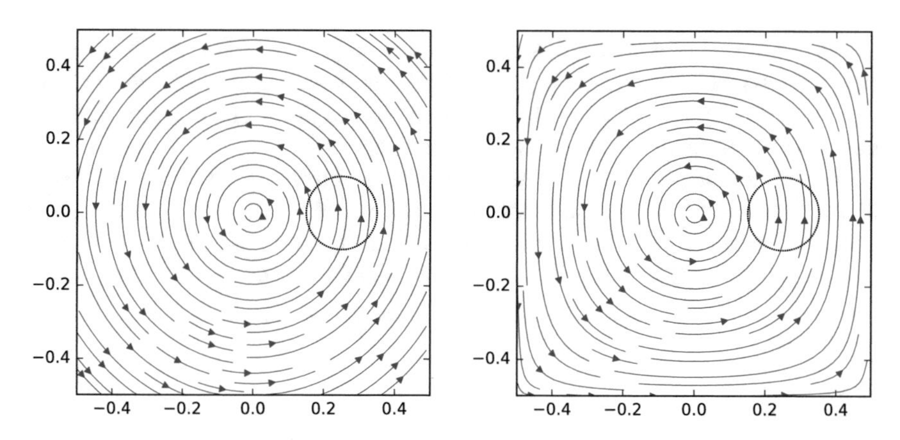

Fig. 7.1 *Left* The *blue flow lines* represent the rigid rotation that we use to test our mesh-to-particles algorithm. The *red dots* represent the initial position of the 100 points to advect. *Right* The setup for the constrained Bell's flow

```
# initialize particle position
centerX=(xMin+xMax)/2.0+xTot*1./4.
centerY=(yMin+yMax)/2.0
radiusCircle=0.1*xTot
dTheta=2*np.pi / pN
for iTheta in np.arange(pN):
    theta=iTheta * dTheta
    px[iTheta]=centerX+np.sin(theta)*radiusCircle
    py[iTheta]=centerY+np.cos(theta)*radiusCircle
```

Here, I have been using a *for* loop only for clarity, and the vectorized implementation is straightforward.

We can now plot the initial conditions using the plotting capabilities of *MatPlotLib*, as shown in Fig. 7.1:

```
plt.streamplot( X[0:nxc], Y[0:nyc], vx[0:nxc,0:nyc].transpose(),
  ↪ vy[0:nxc,0:nyc].transpose(), linewidth=.3)
plt.scatter(px, py, s=.02, alpha=1.0)
plt.show()
```

To have particles that transport material and fields does not entirely solve the problem of diffusion. Instead it transfers the uncertainties that create diffusion into a new uncertainty, which is how to project the information from the mesh to the particles. We have learned in 1.2 that many ways exist to average quantities. We have here four nodes and some particles distributed inside these cells. The easiest way to project the lattice value to each particle is by calculating a weighted average from the four nodes delimiting the cell in which it is located. There are, however, other more sophisticated possibilities, like spectral methods (approximating the field with the summation of sinusoidals), spline development (approximation through delimited polynomials), but these methods go beyond the introductory character of this book and are generally computationally expensive. To use just a linear approach is fast and easier to vectorize and to develop in parallel.

To perform the projection, we need two groups of arrays. One composed by two arrays of integers tracking the index of the cell where the particle is located, here called *cIX* and *cIY*, the other composed by the four weights relating the particles with the four neighboring nodes, called *w1*, *w2*, *w3* and *w4*. Let us allocate them:

```
cIX=np.zeros(pN, int)
cIY=np.zeros(pN, int)
w1=np.zeros(pN, float)
w2=np.zeros(pN, float)
w3=np.zeros(pN, float)
w4=np.zeros(pN, float)
```

We are ready now to run the simulation and test several implementations of the weight and see which one is the best. Since the rotation has an angular velocity of one, the time required for a perfect rotation is π (nondimensional). To observe the solution, it is wise to plot only regular snapshot. For this reason we can create the variable *plotEverySteps*, together with deltaTime that indicates the time necessary for one step:

```
steps=1000
totalTime=2*np.pi
plotEverySteps=20
deltaTime=totalTime/steps
```

7.2.1 Cell-Particles Projections

Since we want to test several weighting strategies, it is smart to create different routines depending on which weighting we like to choose. For example, let us use some standard method used in numerous Particles in Cell implementations such as [76] based on having weights equal to $1/|x_p - x_n|$, i.e., the weight is inversely proportional to the closest node. This solution has the advantage that when a particle is close to a node, a straightforward arithmetic average will give to the particles the node field's value, as illustrated in Fig. 7.2. This feature has, however, also the opposite characteristic, which is to make the projection from the Particles to the Cell unstable, because the node would take the value of the particle just if one particle passes randomly near a node. For this reason a more stable method is recommended for the projecting a field from particles to cell, which is the *Bilinear* scheme, also shown in Fig. 7.2. There are also many more options, some more advanced and sophisticated, like the ones based on *Voronoi* triangulation (e.g., [54]), but their detailed implementation goes beyond the scope of this book.

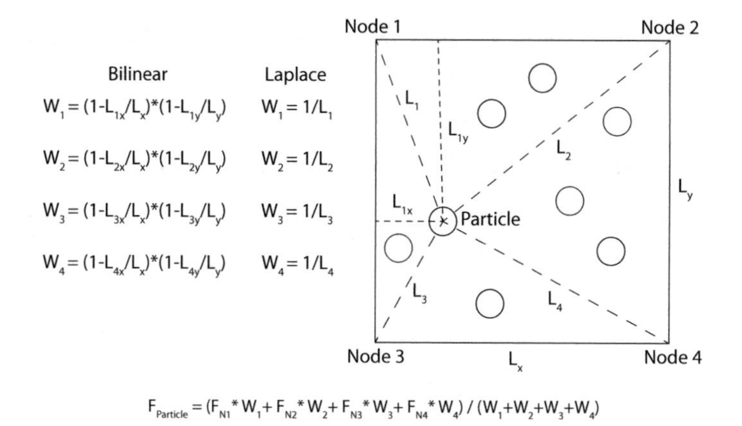

Fig. 7.2 Schematic representation of the weights that relate the particles with the cell nodes. The first *Bilinear* scheme is very stable but can be more diffusive. The second, Laplace is very precise when projecting from the mesh to the particles, because the weight is very high, almost infinite, for a particle very close to a node. It can, however, create instabilities when projecting a field from the nodes to the mesh because a node with take the value of only one particle if this one passes casually very close to a cell corner. Clearly other techniques, more or less nonlinear, exist and the choice depends strongly on the problem to solve

```
from numpy.linalg import norm
def calculateWeightsInverse(xt,yt,X,Y,cIX,cIY,w1,w2,w3,w4,pN):
  # pN is the number of particles
  w1[0:pN] = 1/norm(np.array([xt[0:pN]-X[cIX[0:pN]], yt[0:pN]-Y[cIY[0:pN]]]))
  w2[0:pN] = 1/norm(np.array([xt[0:pN]-X[cIX[0:pN]+1], yt[0:pN]-Y[cIY[0:pN]]]))
  w3[0:pN] = 1/norm(np.array([xt[0:pN]-X[cIX[0:pN]], yt[0:pN]-Y[cIY[0:pN]+1]]))
  w4[0:pN] = 1/norm(np.array([xt[0:pN]-X[cIX[0:pN]+1],
  ↪  yt[0:pN]-Y[cIY[0:pN]+1]]))
  return(w1,w2,w3,w4)
```

The other *Bilinear* implementation instead can be implemented, for example, as:

```
def
calculateWeightsBilinear(xt,yt,X,Y,cIX,cIY,w1,w2,w3,w4,pN,dx,dy):
  w1[0:pN]=(1-(np.abs(xt[0:pN] -X[cIX[0:pN]])/dx))*(1-(np.abs(yt[0:pN]
  ↪  -Y[cIY[0:pN]])/dy))
  w2[0:pN]=(1-(np.abs(xt[0:pN] -X[cIX[0:pN]+1])/dx))*(1-(np.abs(yt[0:pN]
  ↪  -Y[cIY[0:pN]])/dy))
    w3[0:pN]=(1-(np.abs(xt[0:pN] -X[cIX[0:pN]])/dx))*(1-(np.abs(yt[0:pN]
  ↪  -Y[cIY[0:pN]+1])/dy))
    w4[0:pN]=(1-(np.abs(xt[0:pN] -X[cIX[0:pN]+1])/dx))*(1-(np.abs(yt[0:pN]
  ↪  -Y[cIY[0:pN]+1])/dy))
    return (w1,w2,w3,w4)
```

These forms they all require to know the number of particles, but if one is calculating all the weights of all the particles, and not of a subset only, one can use a more readable and compact expression. For example for the arithmetic weights (and an homogeneous mesh) that would be:

```
def
calculateWeightsArithmetic(xt,yt,X,Y,cIX,cIY, w1,w2,w3,w4,pN,dx,dy):
    d=(dx**2+dy**2)**0.5
    w1=1-((xt -X[cIX])**2+(yt -Y[cIY])**2)**0.5/d
    w2=1-((xt -X[cIX+1])**2+(yt -Y[cIY])**2)**0.5/d
    w3=1-((xt -X[cIX])**2+(yt -Y[cIY+1])**2)**0.5/d
    w4=1-((xt -X[cIX+1])**2+(yt -Y[cIY+1])**2)**0.5/d
    return (w1,w2,w3,w4)
```

Based on what we have learned in Chap. 3, these vectorized implementations are extremely fast. In the same way, we can create a routine that projects a field from the mesh to the particles, using a dedicated function. In this first implementation we will project just *velocity* but later we will similarly have any other field:

```
def projectLatticeToParticles(w1,w2,w3,w4,cIX,cIY,f,ft,pN):
    ft[0:pN]=(w1[0:pN]*f[cIX[0:pN],cIY[0:pN]]+
        w2[0:pN]*f[cIX[0:pN]+1,cIY[0:pN]]+
        w3[0:pN]*f[cIX[0:pN],cIY[0:pN]+1]+
        w4[0:pN]*f[cIX[0:pN]+1,cIY[0:pN]+1]) /
  ↪  (w1[0:pN]+w2[0:pN]+w3[0:pN]+w4[0:pN])
    return(ft)
```

where we indicated with *pN* the number of particles that on which the field is projected. Like for the calculation of the weights, if the projection is over all the particles and nodes, this function becomes more compact and readable, given that the index iteration is implicit. But one has to remember that the indexes *cIX* and *cIY* are arrays:

```
def projectLatticeToParticlesCompact(w1,w2,w3,w4,cIX,cIY,f):
    return (w1*f[cIX,cIY] + w2*f[cIX+1,cIY] + w3*f[cIX,cIY+1] +
  ↪  w4*f[cIX+1,cIY+1]) / (w1+w2+w3+w4)
```

7.2.2 Motion of the Particles

We are now ready to simulate the circulation of the particles for rigid rotation. To project the fields to and from the particles we need to find in which *Cell* they are, and do it at every time step, so we need a simple vectorized operation. Since our mesh is regular, we can use a simple proportion. To find *cIX* and *cIY* with a vectorized operation for more sophisticated mesh is usually possible, or eventually a *Cython* routine will always save us. We will see later how also a tree-code implementation can help us for a disordered mesh. In this case we have only:

```
for time in np.arange(steps):
    cIX=((px-xMin)*nxc/xTot).astype(int)
    cIY=((py-yMin)*nyc/yTot).astype(int)

    (w1,w2,w3,w4) = calculateWeightsInverse (px,py,X,Y,cIX,cIY,w1,w2,w3,w4,pN)

    vxt = projectLatticeToParticles (w1,w2,w3,w4,cIX,cIY,vx,vxt,pN)
    vyt = projectLatticeToParticles (w1,w2,w3,w4,cIX,cIY,vy,vyt,pN)

    px+=vxt*deltaTime
    py+=vyt*deltaTime

    if (time % plotEverySteps)==0:
         plt.scatter(px, py, s=.02, alpha=1.0)

plt.xlim(xMin,xMax)
plt.ylim(yMin,yMax)
plt.show()
```

Which produces the output in Fig. 7.3. There, it is also shown the result of a reversed flow. A reversed flow allows in fact to see how great will be cumulative errors. In this case the result is almost perfect, but, this is a product of our very simple rotational flow. A slightly more complicated flow law might show us the weaknesses of this approach, and how to improve it.

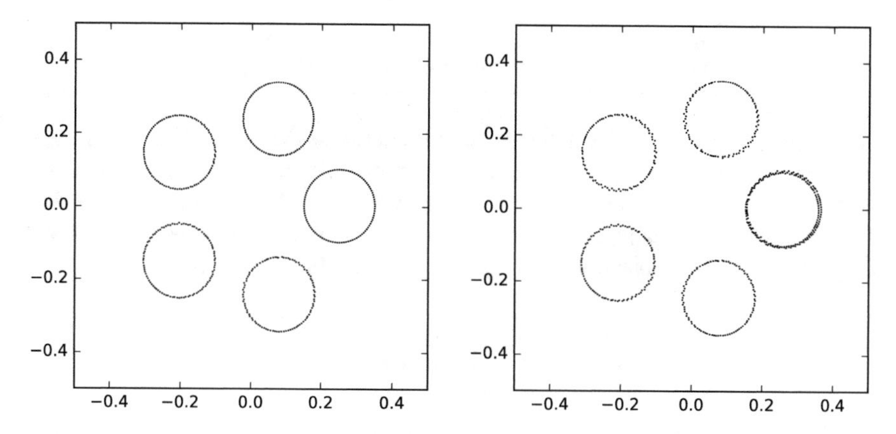

Fig. 7.3 *Left* result of the rigid rotation of a *circle* by using the above mesh-to-particles algorithm. The flow is for a time 2π. *Right* reverse pattern with the same flow. The geometry is almost perfect. For a rotation the interpolation counts very little. Different is the case for a more complex flow

7.3 Thinning Flow

We can now proceed toward testing a more complex field. Computational literature offers a multiplicity of flows with many properties. We use here one introduced by Bell et al. [55], that has been often used in the computational literature to test numerical schemes. It proposes a flow that preserves volumes, as we will verify in another section, but heavily deforms the immersed body, because the intensity of the flow is zero at the edges and at the center, and maximum at the mid distance. In this way it creates heavy stretching and filaments that can be reconstructed by reversing the flow to benchmark the numerical strategy:

$$\begin{aligned} vx &= -\sin^2(\pi x)\sin(\pi y)\cos(\pi y) \\ vy &= \sin^2(\pi y)\sin(\pi x)\cos(\pi x) \end{aligned} \tag{7.5}$$

confined in the box $[0, 1]x[0, 1]$.

Let us use the same algorithm as above, but by replacing the flow with:

```
xNorm=(X+xMin)/xTot
yNorm=(Y+yMin)/yTot
for iy in np.arange(0,nyp):
    vx[:,iy] = np.sin(np.pi*xNorm) * np.sin(np.pi*xNorm) *
  ↪  np.sin(np.pi*yNorm[iy]) * np.cos(np.pi*yNorm[iy])
for ix in np.arange(0,nxp):
    vy[ix,:] =- np.sin(np.pi*yNorm) * np.sin(np.pi*yNorm) *
  ↪  np.sin(np.pi*xNorm[ix]) * np.cos(np.pi*xNorm[ix])
```

in a time T = pi, gives the deformed result shown in the left of Fig. 7.4. A way to check whether this algorithm is accurate in case of such a strong deformation, is to reverse the flow and run it for the same amount of time to see how well it reproduces the initial setup. In order to do so one only needs to change the lines with the lattice-particles projection, and invert the sign of the velocity flow:

```
vxt = projectLatticeToParticles (w1,w2,w3,w4,cIX,cIY,-vx,vxt,pN)
vyt = projectLatticeToParticles (w1,w2,w3,w4,cIX,cIY,-vy,vyt,pN)
```

We have an opportunity now, again, to check the difference between forward, backward and centered integration schemes. We have shown above a forward integration. Let's look for example at how a different scheme would work. One problem of the above scheme is that it uses the *old* velocity, instead of the new one to advect the particle. A much better way to proceed would be to go forward, calculate the new velocity, go back, use the new velocity to go forward. Go forward again, calculate the velocity, go back, an again. This scheme is done in by replacing the python lines:

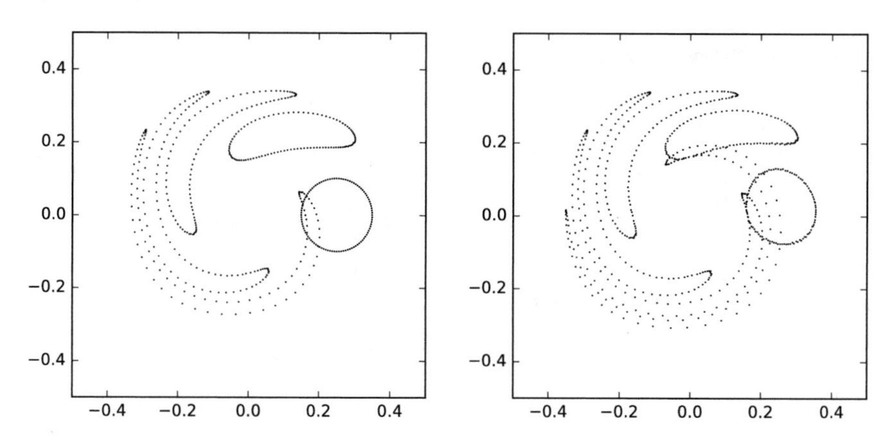

Fig. 7.4 *Left* resulting deformation for the Bell's flow, that stretches the the flow by being characterized by high speed at mid distance between center and domain boundaries. The plotted flows are taken ever 200 steps with maximum time of π (non dimensional). *Right* reverse pattern with the same flow. The geometry is close to good reconstruction, but not perfect

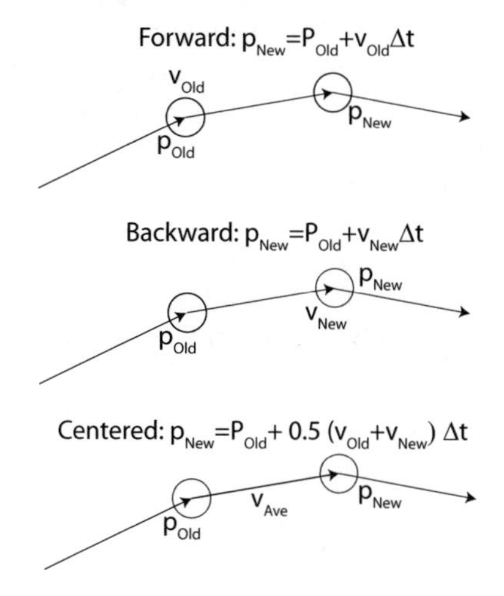

Fig. 7.5 Schematic representation of three main advection schemes for particles: Forward, Backward and Centered. This last is an average of the first two. While the first is the easiest to implement, it also implies greater errors. The other two are more precise, with the centered one being second order stable

```
px+=vxt*deltaTime
py+=vyt*deltaTime
```

with the following ones:

```
pxOld = pxOld + vxt*deltaTime
pyOld = pyOld + vyt*deltaTime
px = pxOld + vxt*deltaTime
py = pyOld + vyt*deltaTime
```

where *pxOld* and *pyOld* are initializated with the same values of *px* and *py*, immediately after their definition:

```
pxOld=px.copy()
pyOld=py.copy()
```

The difference between the two methods is shown in Fig. 7.6. While the advection system above is called *forward*, the other is *backward*, as it uses the velocity of the next step. As one notices none of the two methods is perfect. The forward systematically underestimates the displacement, while the backward scheme overestimates it.

A solution, similar to the schemes that we used in newtonian simulations, is to use a *centered* scheme, where the velocity is averaged between the past and the new step. This can be estimated here for example by calculating the position after a half-step, and use that velocity to transport the particle from the old to the new position. The result of this third approach is also shown in Fig. 7.6, where one notices how superior this approach is. A sketch of the main three advection schemes is in Fig. 7.5. A possible implementation of the centered scheme is the following, in which one calculates the mid points and the velocity at these midpoints, and then uses the velocity at these midpoints to advect the particles starting from the position at the *beginning of the step*:

```
midPx = px + vxp*deltaTime/2.
midPy = py + vyp*deltaTime/2.

cIX=((midPx-xMin)*nxc/xTot).astype(int)
cIY=((midPy-yMin)*nyc/yTot).astype(int)

(w1,w2,w3,w4)=calculateWeights(midPx,midPy,X,Y,cIX,cIY,w1,
w2,w3,w4,nP)

vxp=projectLatticeToParticles(w1,w2,w3,w4,cIX,cIY,vx)
vyp=projectLatticeToParticles(w1,w2,w3,w4,cIX,cIY,vy)

px += vxp*deltaTime
py += vyp*deltaTime
```

where one observes that in order to recalculate the solution one has to determine position of the particles and weights. Although the advantages are well illustrated by the results in Fig. 7.6, it is always important to evaluate whether for our goals it is necessary to take this more complex advection strategy, since it duplicates the calculation necessary to solve our PDE.

More complex approaches are the *leapfrog*, where it is integrated using the a force scheme at half time-steps, and Runge-Kutta of several orders where the information from more intermediate points is used. It is important to take into consideration that these are all *explicit* advection scheme, in the sense that we do not try to iterate to obtain the correct solution. It would be possible in fact to guess the new position, extract the velocity in the new position, return to the initial position and guess again the new position by considering both the initial and final velocity. This approach is however much more computationally demanding and we do not treat it here, for now. Details about many of these explicit schemes are available on many books and online resources so I do not describe them here in detail, but focus on the implementation, that is always similar.

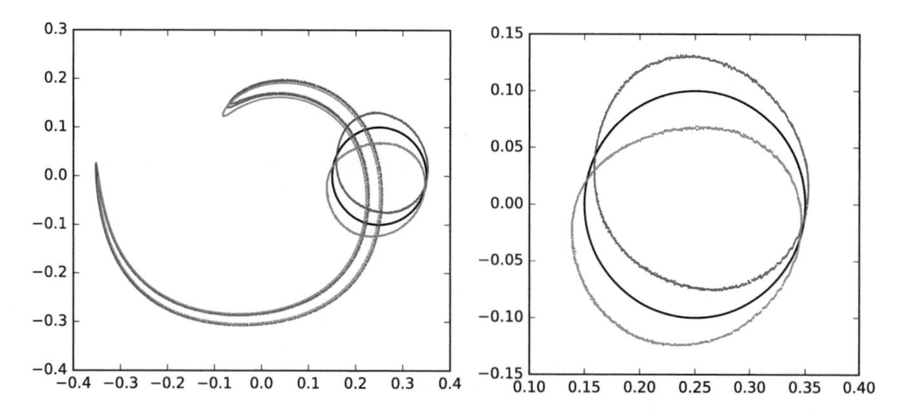

Fig. 7.6 Comparison between three advection schemes: forward (blue), backward (green), centered (red), as sketeched in Fig. 7.5.The test consists in running a full advection until strong thinning of the initial shape is reached, and then reverse the flow for the same amount of time. On the left figure it is shown the last shape after advection, and the last shape after reverse advection. On the right figure the zoom on the final configuration after the reverse flow. The forward and backward schemes fail similarly, one by advecting less, the other more. The centered scheme instead obtains an almost perfect result, as the it is nearly impossible to observe the difference between the initial configuration (in black) and the reconstructed one. The tests have been run with a 1000 timesteps and a lattice mesh of 200×200. It is useful to experiment with different resolutions and number of steps

7.4 Lagrangian Advection of a Continuous Field

The border of a inner domain as the one analyzed above requires a hundred to a thousand particles, and it is perfect for testing the quality of the advection scheme, however for most real problems it will be necessary to advect a continuous field, in general covering the entire domain, sometimes with smooth gradients, others with sharp ones. I will show here how Lagrangian advection can be comfortably and efficiently used for this problem, and later compare this approach with an Eulerian scheme.

This solution requires several steps, among them the distribution of the particles in the entire domain, project a field from the particles to the mesh, project back a field from the mesh to the particles, and also check that step by step the particles to not exit the simulation domain (and eventually take action if they do). In order to do all that the best option is to create separate functions aimed each at one of these scopes.

This is a list of the routines that we need:

A Create and initially deploy the Particles.
B Calculate the Weights (shown in Sect. 7.2).
C Project a continuous field from the Lattice To the Particles (shown in Sect. 7.2).
D Project a continuous field from the Particles To the Lattice
E Constrain the Particle Domain (verify that they don't leave the computing domain).

Two of these routines have been already introduced in Sect. 7.2. Let us focus on the other three.

The first routine is the only one among them that will be called one time (or possibly just few times during the simulation). Once the particles have been deployed, in fact, will be advected using the algorithms described in Sect. 7.2.

In this implementation, the particles are distributed uniformly on the entire domain. In order control their density I introduce a variable P_o, called *particle-Order* in the routine, which indicates how many particles are used to discretize every *side* of each cell. This implies that in each cell there will be P_o^2 particles.

```
def deployParticles(nx,ny,dx,dy,particlesOrder,X,Y):
    particlesPerCell=pPS*pPS
    pN=particlesPerCell*(nx-1)*(ny-1)
    xTemp = X[0]+(np.arange((nx-1)*pPS)+0.5)*dx/pPS
    yTemp = Y[0]+(np.arange((ny-1)*pPS)+0.5)*dy/pPS
    xp, yp = np.meshgrid(xTemp, yTemp)
    xp = xp.reshape(pN)
    yp = yp.reshape(pN)
    return (pN,xp,yp)
```

Here the use of the *reshape* function allows to immediately passing from a *cartesian* mesh to the *serial* set of position of the particles.

During the simulation it can happen that some cells have no particles. This is due to the non-regular distribution of the particles after advection and normal random fluctuations due to the initial position of the particles. This phenomenon is more common for turbulent high Reynolds number flow, however it can happen also for high viscosity variations in Mantle convection, or sharp permeability variations in Darcy flow. The solution to this problem can be quite complex, requiring the redistribution of the position of the particles, which demands a sophisticated algorithm that preserves the information on the field. Attempts to solve this problem for the general case use for example high order splines (e.g. fifth order [83]), but every attempts until present has shown to be problem dependent. In the following simulations, I will only use a large number of particles for each cell since this requires a minimum computing overhead for the simulations that we have here and only for some specific problem that I will indicate the risk to incurring in this difficulty exists.

In most simulations, the boundary conditions should prevent the possibility that the particles may exit from the domain. Still, a numerical round off can always bring a particle slightly outside the main domain. It can also happen that a model precisely predicts that particles enter or exit from the domain. In this case, particles flow in and out of the domain has to be carefully monitored.

I show here an example of a completely vectorized routine that monitors that the particles remain confined within the the four walls of a 2D simulations. The technique is entirely based on the function *np.where()* that we have introduced in Chap. 3. In this version, the particles are redeployed on the boundary where they exited from the domain. In fact, the routine can be rewritten in many ways depending on the modeling goals.

```
def constrainParticleDomain (xp,yp,xMin,xMax,yMin,yMax,dx,dy,particleOrder):
    offsetReinsertParticle = (1./particleOrder)
    xp[xp<xMin] = xMin+offsetReinsertParticle*dx
    yp[yp<yMin] = yMin+offsetReinsertParticle*dy
    xp[xp>xMax] = xMax-offsetReinsertParticle*dx
    yp[yp>yMax] = yMax-offsetReinsertParticle*dy
    return (xp,yp)
```

To run a full simulation, we finally have to write the complementary function to *projectLatticeToParticles* and write another that project a field from the Particles to the Lattice. In effect a fully vectorized Particles to Lattice projection is more complex than the opposite, given that the number of particles in every cell is not known a priori. There are several solution to this problem. The one that I show here is to calculate the weighted field summation for every corner of a cell separately (so, four times), and then calculate the weight summation four times, and finally combine everything together:

```
def projectParticlesToLattice (w1,w2,w3,w4,cIX,cIY,f,fp,plw):
    f[:,:]=0.0
    plw[:,:]=0.0
    f[cIX[:],cIY[:]]+=fp[:]*w1[:]
    f[cIX[:]+1,cIY[:]]+=fp[:]*w2[:]
    f[cIX[:],cIY[:]+1]+=fp[:]*w3[:]
    f[cIX[:]+1,cIY[:]+1]+=fp[:]*w4[:]
    plw[cIX[:],cIY[:]]+=w1[:]
    plw[cIX[:]+1,cIY[:]]+=w2[:]
    plw[cIX[:],cIY[:]+1]+=w3[:]
    plw[cIX[:]+1,cIY[:]+1]+=w4[:]
    f[:,:]/=plw[:,:]
    return (f)
```

Where it is important to consider that *f[]* and *pw[]* have to be allocated before calling this function. The idea was basically to iterate through the particles and not through the nodes of the lattice. This is a simple and practical approach based on the Lagrangian formulation of , however the use of the *for* iteration makes it very slow for practical purposes (i.e. large scale modeling). To solve this problem we need to develop a vectorized version.

A natural technique to do that is to use the *bincount* function of *numpy*, which counts the number of occurrences of each value in an array of non-negative integers. Since *bincount* allows also using weighted summations, it is the perfect tool for our task. Let's look for example at how we can perform an Arithmetic average between all the particles:

```
def projectParticlesToLattice_Arithmetic (w1,w2,w3,w4,cIX,cIY,f,ft,pw):
  (nxp,nyp)=f.shape; nxc=nxp-1; nyc=nyp-1
  f[:,:]=0.; pw[:,:]=0.
  cIG=cIY+nyc*cIX

  f[0:nxc,0:nyc]+=np.bincount(cIG,weights=ft*w1).reshape(nxc,nyc)
  f[1:nxp,0:nyc]+=np.bincount(cIG,weights=ft*w2).reshape(nxc,nyc)
  f[0:nxc,1:nyp]+=np.bincount(cIG,weights=ft*w3).reshape(nxc,nyc)
  f[1:nxp,1:nyp]+=np.bincount(cIG,weights=ft*w4).reshape(nxc,nyc)
  pw[0:nxc,0:nyc]+=np.bincount(cIG,weights=w1).reshape(nxc,nyc)
  pw[1:nxp,0:nyc]+=np.bincount(cIG,weights=w2).reshape(nxc,nyc)
  pw[0:nxc,1:nyp]+=np.bincount(cIG,weights=w3).reshape(nxc,nyc)
  pw[1:nxp,1:nyp]+=np.bincount(cIG,weights=w4).reshape(nxc,nyc)

  f[:,:]/=pw[:,:]
return (f)
```

Similarly one can develop an implementation that uses a Harmonic average:

```
def projectParticlesToLattice_Harmonic (w1,w2,w3,w4,cIX,cIY,f,ft,pw):
  (nxp,nyp)=f.shape; nxc=nxp-1; nyc=nyp-1
  f[:,:]=0.0; pw[:,:]=0.0
  cIG=cIY+nyc*cIX

  f[0:nxc,0:nyc]+=np.bincount(cIG,weights=w1/ft).reshape(nxc,nyc)
  f[1:nxp,0:nyc]+=np.bincount(cIG,weights=w2/ft).reshape(nxc,nyc)
  f[0:nxc,1:nyp]+=np.bincount(cIG,weights=w3/ft).reshape(nxc,nyc)
  f[1:nxp,1:nyp]+=np.bincount(cIG,weights=w4/ft).reshape(nxc,nyc)
  pw[0:nxc,0:nyc]+=np.bincount(cIG,weights=w1).reshape(nxc,nyc)
  pw[1:nxp,0:nyc]+=np.bincount(cIG,weights=w2).reshape(nxc,nyc)
  pw[0:nxc,1:nyp]+=np.bincount(cIG,weights=w3).reshape(nxc,nyc)
  pw[1:nxp,1:nyp]+=np.bincount(cIG,weights=w4).reshape(nxc,nyc)

  f[:,:]=pw[:,:]/f[:,:]
return (f)
```

Where this last implementation requires that the function in the Harmonic average has never to reach zero. Typically this kind of average is used to calculate average viscosity in system where it varies very strongly.

The most important advantage of this approach is clearly that the operation is now entirely vectorized, and that it can be naturally split among in subsets of particles, and therefore immediately written in parallel. Naturally the summation has to be divided in chunks of the length of an exact fraction of the number of particles.

Let's look now at how all this can be put together in one simulation. We will create a bell-shaped temperature anomaly in one corner of the domain and then advect with the Bell's flow that we used above. I assume here the same parameters and flow as in the above simulations, then the new lines of code are:

```
nxp=51; nxc=nxp-1; xTot=1.0; dx=xTot/nxc; xMin=-xTot/2.0; xMax=xTot/2.0
nyp=51; nyc=nyp-1; yTot=1.0; dy=yTot/nyc; yMin=-yTot/2.0; yMax=yTot/2.0
pPS = 3 # particles per side of the cell

X=np.arange(nxp)*dx+xMin
Y=np.arange(nyp)*dy+yMin
vx=np.zeros((nxp,nyp), float)
vy=np.zeros((nxp,nyp), float)

plw = np.zeros((nxp,nyp), float) #particles -> lattice weights

# initialize particle position
(nP,px,py)=deployParticles(nxp,nyp,dx,dy,pPS,X,Y)
# allocate the arrays vxp, vyp, cIX, cIY, w1, w2, w3, w4

# initialize the flow:
xNorm=(X+xMin)/xTot; yNorm=(Y+yMin)/yTot
for iy in np.arange(0,nyp):
    vx[:,iy]=np.sin(np.pi*xNorm)*np.sin (np.pi*xNorm)*np.sin(np.pi*yNorm[iy])
*np.cos(np.pi*yNorm[iy])
for ix in np.arange(0,nxp):
    vy[ix,:]=-np.sin(np.pi*yNorm)*np.sin (np.pi*yNorm)*np.sin(np.pi*xNorm[ix])
*np.cos(np.pi*xNorm[ix])

# initialize the temperature field on the particles with a bell shape centered
  ↪ in xTot/4, yTot/4

T=np.zeros((nxp,nyp), float) #mesh
Tt=np.exp(-(px-xTot/4)**2*50/xTot**2)
Tt*=np.exp(-(py-yTot/4)**2*50/yTot**2)
steps=101;totalTime=2*np.pi/2; deltaTime=totalTime/steps

# the simulation begins
for time in np.arange(steps):
    cIX = ((px-xMin)*nxc/xTot).astype(int)
    cIY = ((py-yMin)*nyc/yTot).astype(int)
    (w1,w2,w3,w4) =
calculateWeightsBilinear(px,py,X,Y,cIX,cIY,w1,w2,w3,w4,nP,dx,dy)

    # projection of the velocity and the field from the lattice to the
  ↪ particles
    T=projectParticlesToLattice_Arithmetic (w1,w2,w3,w4,cIX,cIY,T,Tt,plw)
    vxp = projectLatticeToParticles(w1,w2,w3,w4,cIX,cIY,vx)
    vyp = projectLatticeToParticles(w1,w2,w3,w4,cIX,cIY,vy)

    # explicit calculation of the of the new position using a mid step
    midPx = px + vxp*deltaTime/2.
    midPy = py + vyp*deltaTime/2.
```

```
    cIX=((midPx-xMin)*nxc/xTot).astype(int)
    cIY=((midPy-yMin)*nyc/yTot).astype(int)
    (w1,w2,w3,w4)=calculateWeightsBilinear (midPx,midPy,X,Y,cIX,cIY,w1,w2
,w3,w4,pN,dx,dy)
    vxp=projectLatticeToParticles (w1,w2,w3,w4,cIX,cIY,direction*vx)
    vyp=projectLatticeToParticles (w1,w2,w3,w4,cIX,cIY,direction*vy)
    px += vxp*deltaTime
    py += vyp*deltaTime

# monitor that the particles remain in the computing domain
    (px,py) = constrainParticleDomain(px,py,xMin,xMax,yMin,yMax,dx,dy,pPS)
```

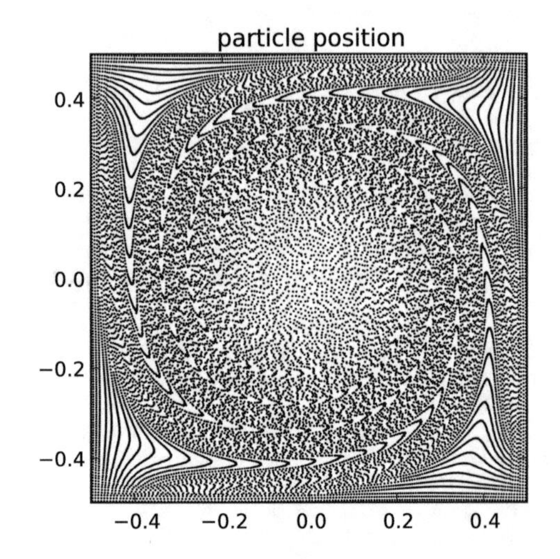

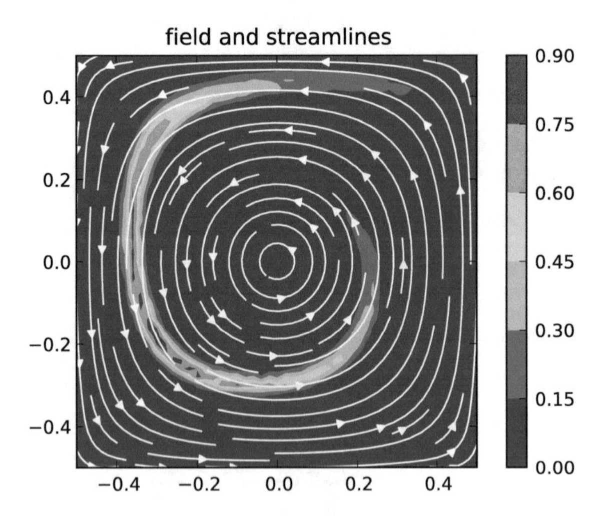

Fig. 7.7 *Left* position of the particles after 100 steps for a 50×50 regular mesh and 3×3 particles per cell. The distribution shows how in some regions, corresponding to where the speed is greatest, the distribution of particles is poorer, and in others the particles density is greater. *Right* the calculated advected field is shown above the flow field. The anomaly is advected more where the density

I show the position of the particles after 100 steps and the advected temperature field in Fig. 7.7. I show here a relatively low resolution model (50 x 50) in order to emphasize how the distribution of particles is strongly irregular, history-dependent and flow-law-dependent. When the field is projected from only few particles to a cell strong fluctuations appear, as it is observed but the lack of continuity of the red region in the flow (on the right of the figure). One can run this same simulation with a mesh of 100 x 100 and eliminate this imperfection. One can try also to duplicate the parameter P_o but that is not as much beneficial.

As we will see in the next chapters particles advection is not the most computationally demanding task in a modeling simulations, as to solve the fundamental momentum equations is much harder, particularly when it is non-homogeneous or non-linear. Problem 7.4 proposes some tests to check computationally efficiency of each routine and also the scaling with problem size. Problem 7.5 instead is about parallelization of the advection routine. Since all the routines are vectorized, a parallel version can be in general developed in a straightforward way, at least for few hundreds of processors.

7.5 Upwind Scheme Versus Lagrangian Transport

There are cases in which such a sophisticated method as Lagrangian advection is not necessary, the first being the standard thermally driven convection. When gradients are not very steep a more convenient and less memory and computing demanding algorithm is possible. Let us look now at how a less accurate but much faster, less memory consuming and always balanced approach can be written and simply vectorized.

The most well-known algorithm is called the 'Upwind scheme' and is based on the simple idea that every total derivative respect to space of a field of a moving fluid can be expressed as the product of the velocity field for the gradient of the field itself:

$$\frac{dF}{dt} = \frac{\partial F}{\partial t} + \frac{\partial x_i}{\partial t}\frac{\partial F}{\partial x_i} = \frac{\partial F}{\partial t} + v_i \frac{\partial F}{\partial x_i} \tag{7.6}$$

The terms on the right to the Eulerian derivative $\frac{\partial F}{\partial t}$ are structured as a velocity for gradient, therefore in the discretized lattice scheme it is more correct to assume a biased location of the gradient on the side in which the fluid is flowing. This simple technique is very ancient and was already introduced by Courant and co-workers [9]. Its stability is also defined by the so called Courant criterium $v\delta t/\Delta x < 1$. While the advantage of this approach is clearly speed and minimized memory usage (compared to using particles) it is subjected to quite a substantial diffusion of sharp gradients, which makes its use inappropriate for accurate tracking of material boundaries.

A modern version of this algorithm is the so-called *Donor-Cell algorithm*. Its goal is the same, but it is based on fluxes, therefore its physical interpretation is more clear. When the material flows from left to right ($v_x > 0$), then the flux of the field F through the interface between one cell and the next is $F_{leftCell}\dot{v}_x^{left} - F_{Cell}\dot{v}_x^{right}$, while if $v_x < 0$ the flux is $F_{Cell}\dot{v}_x^{right} - F_{rightCell}\dot{v}_x^{left}$.

To vectorize the donor-cell method the simplest approach is to calculate the flux through the each cell and then use a boolean index and its negative (not index) to select which direction for the flow needs to be taken. This is done for every axis separately. It follows here an example in 2D.

```
def vectorizedUpwind(F,vx,vy,dx,dy,dt):
    (nx,ny) = F.shape
    dFplus = np.zeros((nx,ny),float)
    dFminus = np.zeros((nx,ny),float)
    dF_dt = np.zeros((nx,ny),float)

    ind = vx>0.0  # flag array for x upwind
    ind[0,:] = True; ind[nx-1,:]=False

    dFplus[1:-1,:] = (F[:-2,:]*vx[1:-1,:] - F[1:-1,:]*vx[2:,:])/dx
    dFminus[1:-1,:] = (F[1:-1,:]*vx[1:-1,:] - F[2:,:]*vx[2:,:])/dx
    dF_dt = (ind * dFplus + ~ind * dFminus)

    ind = vy>0.0  # flag array for y upwind
    ind[:,0] = True; ind[:,ny-1]=False

    dFplus[:,1:-1] = (F[:,:-2]*vy[:,1:-1] - F[:,1:-1]*vy[:,2:])/dy
    dFminus[:,1:-1] = (F[:,1:-1]*vy[:,1:-1] - F[:,2:]*vy[:,2:])/dy
```

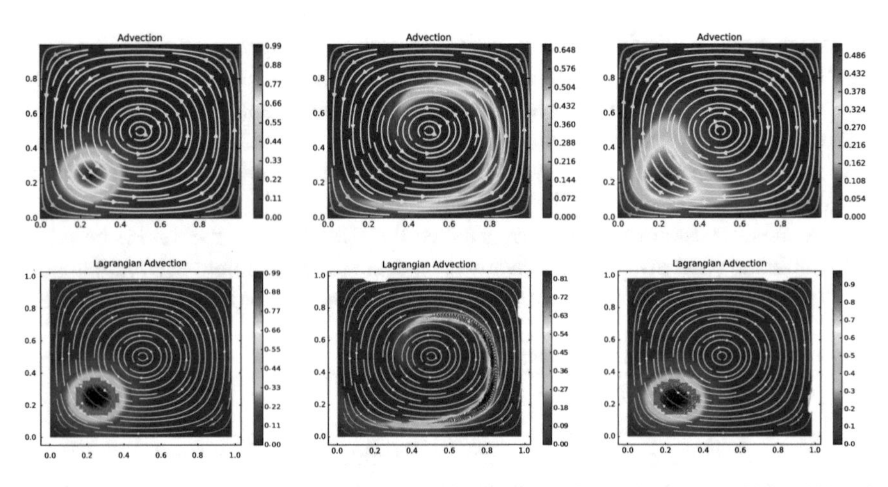

Fig. 7.8 *Top* Simulation of Upwind advection using the Bell's flow. Resolution is 200 × 200 and the number of timesteps is $n = 500$. *Left* initial conditions of the field anomaly. The streamlines are shown in the background. *Middle* final configuration after running a forward advection for a time $\Delta t = 2.0$. *Right* configuration after running a backward advection with the same flow field, opposite direction, for the same time. One observes that the center of the anomaly is symmetric and placed at the same position at the corner, however diffusion is strong along the flow. *Bottom* Same simulation but using Lagrangian particles to advect the flow. The cartesian resolution is here only 50 × 50, same number of steps and 4 × 4 particles for each cell. The position of the particles is shown with a black dot only for the particles that correspond to value of more than $(T_{max} + T_{min})/2$. One observes that the backward flow is non-diffusive and centered. Lagrangian transport is therefore superior even with a less resolved cartesian grid

```
    dF_dt += (ind * dFplus + ~ind * dFminus)

    return (F + dt * dF_dt)
```

The Fig. 7.8 shows how the upwind scheme is well balanced, without bias, but how the backward simulation displays diffusion, in particular in the direction of the flow. We therefore conclude that the upwind scheme is conservative and balance, it requires little memory, but it is very diffusive. In Fig. 7.8 I compare the result of the Upwind scheme with the Lagrangian advection introduced in this chapter where it is clear that the Lagrangian approach is non-diffusive, but more complex to implement and more memory demanding.

Summary

- Two standard flow laws are introduced: rigid rotation and a thinning flow in a box. They are also used in the next chapters to calculate other properties.
- Three techniques to calculate the weights to project continuous fields between mesh and particles have been introduced based on different averaging techniques.
  - arithmetic
  - bilinear
  - inverse
- Several techniques to advect particles in a cartesian mesh are illustrated. It is shown how important is to calculate the mid-value of the velocity.
- It is shown how to project a field from the particles to the mesh. To vectorize this operation a summation that iterates over the particles has to be introduced.
- Particles are displaced and some cells increase the number of particles, other have few particles. This has to be monitored and handled.
- Particles advection routines can be all entirely vectorized, which makes them easily to implement with NumPy and straightforward to parallelize.

Several of the functions introduced in this chapter will be employed again when running more advanced routines, in the next sections. I recommend to create a personal little library where the benchmarked functions are stored and can be quickly later recalled.

Problems

7.1 Use the other two weights (Arithmetic and Bilinear) to test the quality of the advection using the Bell's flow as described in Sect. 7.3. By observing closely the point spacing, you can determine how smooth each algorithm is. Try to quantify this effect and develop a feeling for how these schemes affect the numerical result.

7.2 The advection algorithm described in Sect. 7.3 uses constant spacing. Can you write the functions for calculating the advection always on a Cartesian grid but with generic X and Y seedings? You have to rewrite mainly how to calculate in which cell you are, and find an efficient way to do it.

7.3 One alternative to Bell's flow for testing a tracking algorithm was introduced by Rider and Kothe, 1995 [79] and is now commonly used. The flow is represented by the two simple expressions:

$$
\begin{aligned}
vx &= \cos(\pi x)\sin(\pi y) \\
vy &= -\sin(\pi x)\cos(\pi y)
\end{aligned}
\tag{7.7}
$$

Run the tests introduced in this chapter and verify whether the tracking algorithm still works with it and how the upwind scheme performs.

7.4 There are several possibly computationally demanding operations in calculating the particles advection with a large number of particles. Using the iPython magic keyword *%timeit* check the speed necessary to run the particles to cell projection (projectParticlesToLattice), cell to particles projection (projectLatticeToParticles), the particle monitoring routine (constrainParticleDomain) and the calculation of the weights (calculateWeightsArithmetic), as described in Sect. 7.4. With some testing, by increasing the problem resolution, notice which one dominates large problems and check whether they all scale linearly, which is the optimal scenario.

7.5 Particles advection is in general very straightforward to parallelize as the routines that are called use only vectorized operations. As an exercise write a parallel version for the most demanding routines, after having tested them in Problem 7.4 of Sect. 7.4. Start with large problems, like 200×200 and above, where the parallelization is most effective.

Chapter 8
Operator Formulation

"Operator! Give me the number for 911!"—Homer Simpson

Abstract This chapter focuses on creating operators made of matrices that calculate derivatives. These derivative operators are in turn used to calculate the divergence of a field and the strain rates. The divergence is then used to prove the continuity of the flow introduced in the past chapter. It shown in particular the need of calculating derivatives centered in the same location (e.g., cell center) or a non-negligible systematic error will appear. It is also illustrated how the formulation of the average and centering formulations for the operators depends on cell ordering.

We start now a travel into the world of linear operators. This is probably the most abstract concept in this book, therefore it will require some effort from the reader who is not familiar with matrices and vectors. As for the rest of this book, we learn this subject by a series of examples, which should allow you to develop a general understanding of how to create a linear operator in the form of a two-dimensional array in Numerical Python. We start with calculating a Strain Rate tensor operator that can apply to any predefined flow law. Then we will use it to create an operator that verify the entire continuity equation for an incompressible flow. We already introduced the continuity equation in Chap. 6, however now we will see how it emerges from the Strain Rate Tensor Operator. In fact, the continuity equation corresponds to setting the trace of the Strain Rate tensor to zero. A sketch representing the continuity equation is shown in Fig. 8.1.

$$\Delta j_x(x_0, y_0) = j_x\left(x_0 + \frac{\Delta x}{2}, y_0\right) - j_x\left(x_0 - \frac{\Delta x}{2}, y_0\right) \approx \frac{\partial j_x}{\partial x}(x_0, y_0)\,\Delta x \quad (8.1)$$

by extending it in the other directions, one can write the entire continuity equation in function of the density ρ and of the flux $\mathbf{j}$:

G. Morra, *Pythonic Geodynamics*, Lecture Notes in Earth System Sciences,
DOI 10.1007/978-3-319-55682-6_8

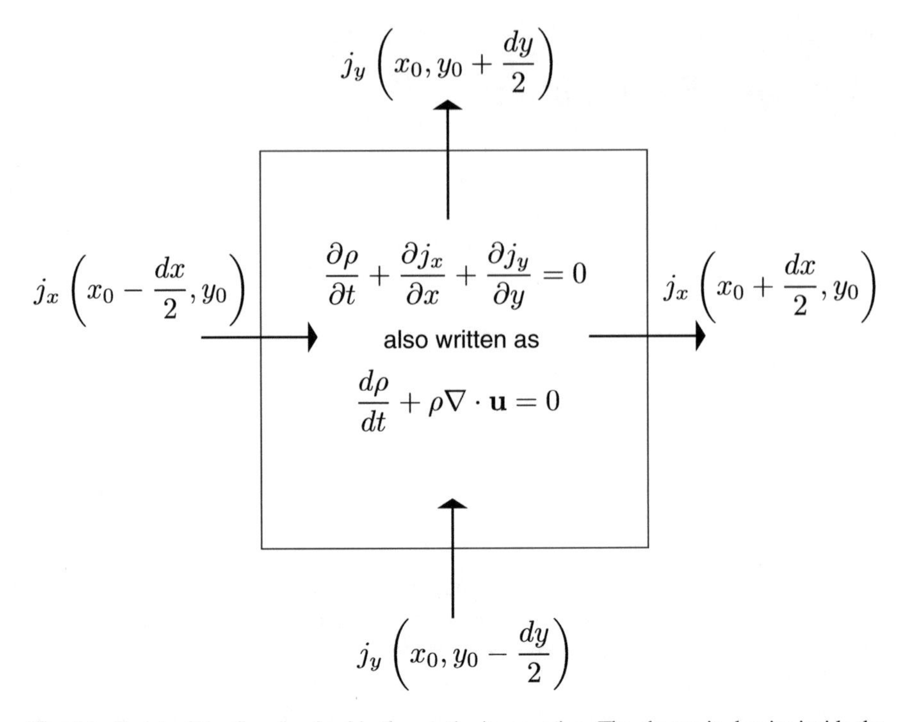

Fig. 8.1 Sketch of the flow involved in the continuity equation. The change in density inside the Eulerian volume is the net difference between the incoming and outgoing fluxes. The result is the continuity equation, that can be either expressed in terms of density ρ and fluxes $\mathbf{j}$, or in function of ρ and the velocity $\mathbf{v}$. Fluxes are calculated as average on each side. A first order Finite Difference representation is shown in 8.2

$$\frac{\partial \rho}{\partial t} + \frac{\partial j_x}{\partial x} + \frac{\partial j_y}{\partial y} = \frac{\partial \rho}{\partial t} + \nabla \cdot \mathbf{j} = 0 \tag{8.2}$$

This can be rewritten in function of the density ρ and of the velocity $\mathbf{v}$ by exploiting the definition of flux and calculating the derivative in detail:

$$\nabla \cdot \mathbf{j} = \nabla \cdot (\rho \mathbf{v}) = (\nabla \rho) \cdot \mathbf{v} + \rho \nabla \cdot \mathbf{v} \tag{8.3}$$

where with $\nabla \rho$ we expressed the gradient vector of the density $\left(\frac{\partial \rho}{\partial x}, \frac{\partial \rho}{\partial y}\right)$. Finally this allows rewriting the continuity equation as:

$$\frac{d\rho}{dt} + \rho \nabla \cdot \mathbf{v} = 0 \tag{8.4}$$

where we used the definition of the full derivative $\frac{d\rho}{dt} = \frac{\partial \rho}{\partial t} + (\nabla \rho) \cdot \mathbf{v}$. The most important consequence of this expression is that for the special case of incompressible flow $\frac{d\rho}{dt} = 0$, and therefore continuity equation simply reduces to

$$\nabla \cdot \mathbf{v} = 0 \tag{8.5}$$

In this chapter we will verify this simplified version of the continuity equation for the flows introduced in Chap. 7, and more in general calculate the full strain rate tensor. Since strain rates involve both the derivatives in the x and y directions of fluxes in x and y directions, this will give us the opportunity to understand how to calculate a two-dimensional derivative operator. The 3D extension follows in a straightforward manner. In particular, it will be important to show how determined choices only allow to build derivative operators that are correct to machine precision even with modest model resolutions.

8.1 Strain Rates

Given an internally deforming material in a box, how do we quantify the rate and the amount of deformation? Since every point of the body is moving at a different velocity (speed and direction) the deformation is physically quantified by how different is the displacement (**s**) or velocity (**v**) from one point to another. By looking at the limit to infinitely close points, we obtain two quantities called *Strain* ε and *Strain Rate* $\dot{\varepsilon}$:

$$\begin{aligned} \varepsilon_{ij} &= \frac{1}{2}\left(\frac{\partial s_i}{\partial x_j} + \frac{\partial s_j}{\partial x_i}\right) \\ \dot{\varepsilon}_{ij} &= \frac{1}{2}\left(\frac{\partial v_i}{\partial x_j} + \frac{\partial v_j}{\partial x_i}\right) \end{aligned} \tag{8.6}$$

To understand strain, strain rate, and the associated energy, the best is to use an example. We can take the 2D flow in Eq. (7.7), and calculate its x and y derivatives, which will give us the strain rate. There are fundamentally two strategies to calculate derivatives using Numerical Python. One is to do it directly, by differentiating the array with the definition of derivatives, the other is to create an *operator* in form of a matrix that calculates the derivatives for us. Let us look at both approaches, since they are both useful for achieving different targets.

The four analytical derivatives of the thinning flow that we introduced above ($vx = -\sin^2(\pi x)\sin(\pi y)\cos(\pi y)$ and $vy = \sin^2(\pi y)\sin(\pi x)\cos(\pi x)$) allows us to calculate the exact strain rate:

$$\dot{\varepsilon} = \begin{pmatrix} \frac{\partial v_x}{\partial x} & \frac{\partial v_x}{\partial y} \\ \frac{\partial v_y}{\partial x} & \frac{\partial v_y}{\partial y} \end{pmatrix} = \begin{pmatrix} -2\pi \sin(\pi x)\cos(\pi x)\sin(\pi y)\cos(\pi y) & -\pi \sin^2(\pi x)[\cos^2(\pi y) - \sin^2(\pi y)] \\ \pi \sin^2(\pi y)[\cos^2(\pi x) - \sin^2(\pi x)] & 2\pi \sin(\pi y)\cos(\pi y)\sin(\pi x)\cos(\pi x) \end{pmatrix}$$

A quick look at the analytically calculated strain rates tells us that this flow is conservative, i.e., $\frac{\partial v_x}{\partial x} + \frac{\partial v_y}{\partial y} = 0$. Let us now check whether this property is found by our calculation of the derivatives.

The most important fact about calculating a derivative on a discretized lattice is that we have to think about where it is calculated. If we take the values at the nodes of a lattice, the derivatives will be calculated at the midpoint between two nodes. This implies that the x-derivative and the y-derivative will be located in two different points. Let us see what are the implications of such an approximation by testing the validity of the continuity equation in the most straightforward way:

```
Dvx = (Vx[1:nxp,0:nyp-1]-Vx[0:nxp-1,0:nyp-1])/dx
Dvy = (Vy[0:nxp-1,1:nyp]-Vy[0:nxp-1,0:nyp-1])/dy
p = plt.imshow(DVx+DVy)
plt.colorbar()
plt.show()
```

Where Vx, Vy are the same defined in the past chapter. *Dvx* and *Dvy* do not need to be allocated. By defining them in this way, the arrays are automatically allocated by the Python parser, as which defined them as $(nxp - 1) * (nyp - 1) = nxc * nyc$ float arrays. This method is very compact and fast, however the result, shown on the left of Fig. 8.3, is not satisfactory. The intensity of the flow is about 1, and the precision of the calculation of the continuity equation oscillates between -0.05 and $+0.05$ as.

Clearly, this performance can be improved by increasing the mesh resolution. One can in fact calculate it with a 1000×1000 mesh and indeed it improves of one order of magnitude. But why is this happening? Can we improve this calculation without increasing the resolution? Indeed there is a better way to improve this result, which is to calculate the two derivatives at the same point, the center of the cell, as shown in Fig. 8.2. Let us do it numerically:

```
vxSideX = 0.5*(vx[:,0:nyp-1]+vx[:,1:nyp])
vySideY = 0.5*(vy[0:nxp-1,0:nyp]+vy[1:nxp,0:nyp])
Dvx=( vxSideX[1:nxp,:] - vxSideX[0:nxp-1,:] )/dx
Dvy=( vySideY[:,1:nyp] - vySideY[:,0:nyp-1] )/dy
```

whose plot is shown on the right of Fig. 8.3. We see that the difference is striking, that the cell-centered approximated equation is many orders of magnitude better than the approximation that we made earlier. In fact it simply reaches machine precision for this perfectly conservative flow.

This very important lesson instructs us on the importance of using rigorously cell-centered finite difference. This technique, analogue to staggered-grid finite difference (where the staggered-grid is a half side shifted grid, with the nodes at the center of the standard grid), to finite volume and and others, is based on the technical, but essential, property that the physical equations are solved at the center of the cell.

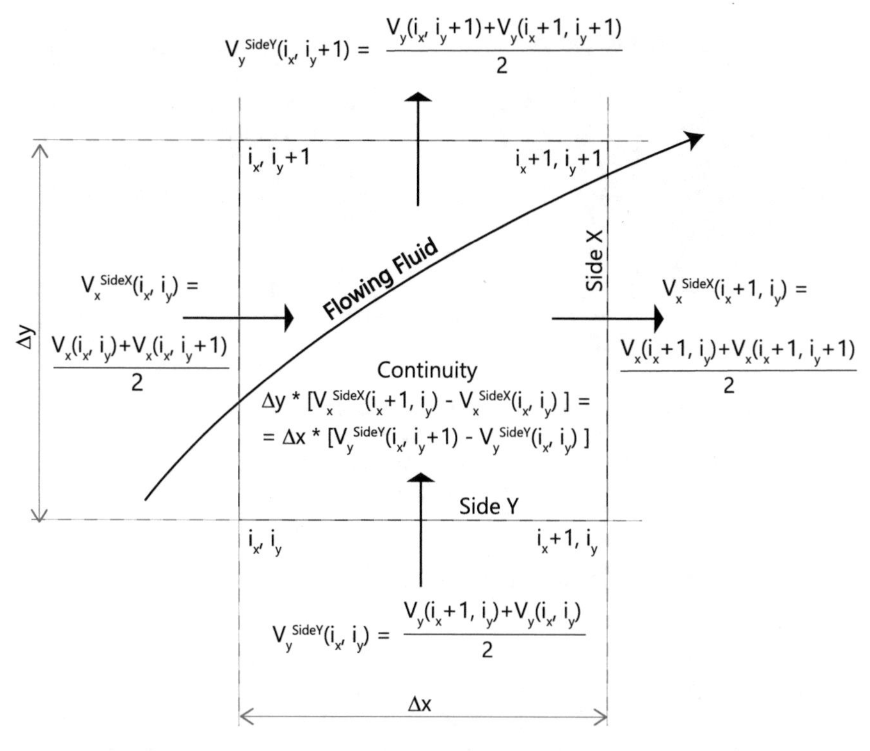

Fig. 8.2 Representation of the calculation of the flow entering and leaving one cell. Conservation of volume requires that the total flux be zero. Every *arrow* represents the linear average flow on that side. The velocity on that side is calculated as the average on that side

The implication is that we need to develop algorithms such that all the quantities are finally calculated at the same point, most commonly the center of the cell.

Let us finally test our cell-centered approximation, calculate the entire strain rate at the cell center and compare it with the analytical one. We have already seen how to calculate $\dot{\varepsilon}_{xx}$ and $\dot{\varepsilon}_{yy}$, by averaging nodal velocity on one axis, and then differentiating respect to the other axis, in order to end up in the cell center. In order to calculate $\dot{\varepsilon}_{xy}$, that is equal by definition to $\dot{\varepsilon}_{yx}$, we need to average on the alternative axis to the one in which we have to derive. This operation is illustrated in detail in Fig. 8.4. Following this approach the calculation of $\dot{\varepsilon}_{xy}$ simply reduces to:

```
vySideX = 0.5*(vy[:,0:nyp-1]+vy[:,1:nyp])
vxSideY = 0.5*(vx[0:nxp-1,:]+vx[1:nxp,:])
Exy = 0.5*( vySideX[1:nxp,:] - vySideX[0:nxp-1,:] )/dx + 0.5*( vxSideY[:,1:nyp]
  ↪  - vxSideY[:,0:nyp-1] )/dy
```

Which, it can be verified, is in excellent agreement with the analytical solution.

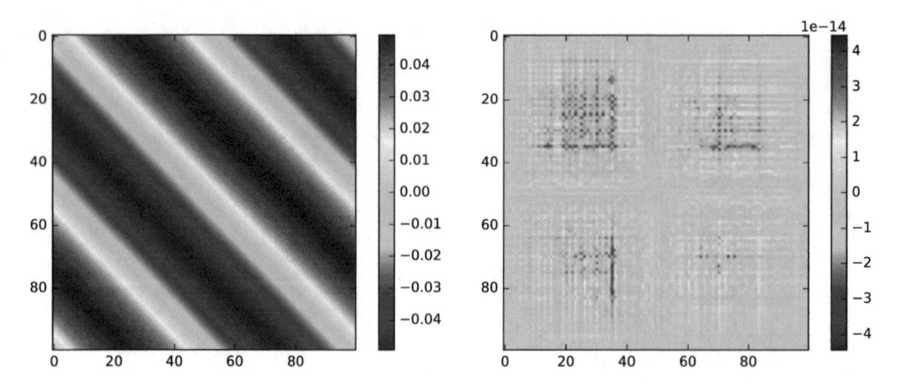

Fig. 8.3 *Left* result of the continuity equation of the flow expressed by Eq. (7.7) when the derivatives are calculated from the values at the nodes. *Right* the same but the value is now calculated on the faces of each cell, and their value is used for the derivation. In this way the two derivatives are calculated at the center of the cell, as shown in Fig. 8.2

8.2 Cell-Centered Strain Rates from Linear Operators

Until now, we have calculated derivatives explicitly. In the future, however, we plan to solve more complex problems where, in general, the law is not known. For example, in the next chapters we will calculate the evolution of a system knowing only its physical state (density, viscosity, forces, etc.), and solve the flow field from the momentum and continuity equations only. In order to develop algorithms able to solve these types of systems we will take advantage of using *Operators*.

Operators are functional objects that act on functions and perform operations on them. For example, a derivative is an operation, and we will create operators derivatives to do that in different context. The advantage of creating these operators is that it can be shown that for any type of discretization of the space, they can be described by a matrix. And also that they form a so-called *Group*, which in simple terms means that we can apply an operator to another operator and obtain a new operator. For example, we can apply the derivative operator to another derivative operators and obtain a second derivative operator. In the same way, we can create a *Divergence Operator*, and so on. These operations are all permitted because these operators, in 1D, 2D, or 3D, are always 2D arrays (matrices) that operate on (1d) arrays, since we can always *reshape* the discretized field that we are interested in a 1D array, as seen in the past section.

Furthermore, operations like the calculation of the momentum can be inverted if performed with linear operators and if these operators have the right properties. Let us see how can we write a linear operator that does the same calculations done in Sect. 8.1. First of all, we have to transform our 2D field array in a 1D array. This is done automatically by NumPy using the function *reshape*. Since we have ordered our field as x, y, *reshape* will automatically reorder any array first along the y axis and then the x axis. Using this property we can create a function that creates an operator

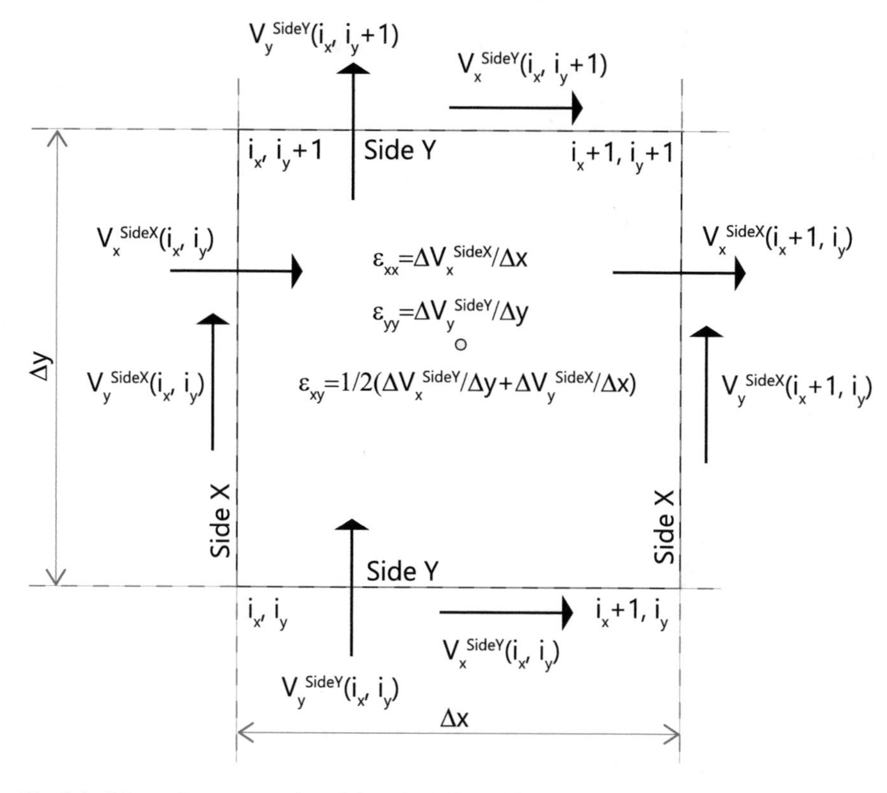

Fig. 8.4 Schematic representation of the calculation of the strain rates xx, yy and xy for a generic cell. In order to obtain always the strain rate at the center of the cell, it is necessary to properly choose between averaging the velocity on the X or Y side. The side velocity is the average of the node velocity at the edges of the side

which takes a $nxp \times nyp$ array and returns a $nxp \times nyp - 1$ array where the values have been averaged on the *side X* of each cell, i.e., along the Y axis:

```
def buildSideOperatorX(nxp,nyp):
  nyc=nyp-1
  block=np.zeros((nyc,nyp),float);
  block[0:nyc,0:nyc]+=0.5*np.diag(np.ones(nyc),0)
  block[0:nyc,1:nyp]+=0.5*np.diag(np.ones(nyc),0)
  return (np.kron(np.identity(nxp),block))
```

where only vectorized operations were used. Here the strategy is to create a block for each line that is *nyp* long, and then use the function *kron* to place this block *nxp* times along the diagonal. It is very important to realize that since the block is not squared, but $(nyp - 1) \times nyp$, the final operator will not be squared. This is normal since the goal is to reduce the number of points where the field is calculated. To show how this operator looks like, I show here the case for nxp = 3 and nyp = 4, where one can see the jump from one block to the next every three rows, and the different number of total lines and rows:

```
[[ 0.5  0.5  0.   0.   0.   0.   0.   0.   0.   0.   0.   0. ]
 [ 0.   0.5  0.5  0.   0.   0.   0.   0.   0.   0.   0.   0. ]
 [ 0.   0.   0.5  0.5  0.   0.   0.   0.   0.   0.   0.   0. ]
 [ 0.   0.   0.   0.   0.5  0.5  0.   0.   0.   0.   0.   0. ]
 [ 0.   0.   0.   0.   0.   0.5  0.5  0.   0.   0.   0.   0. ]
 [ 0.   0.   0.   0.   0.   0.   0.5  0.5  0.   0.   0.   0. ]
 [ 0.   0.   0.   0.   0.   0.   0.   0.   0.5  0.5  0.   0. ]
 [ 0.   0.   0.   0.   0.   0.   0.   0.   0.   0.5  0.5  0. ]
 [ 0.   0.   0.   0.   0.   0.   0.   0.   0.   0.   0.5  0.5]]
```

Similarly an operator can be build for the side Y, in the X axis direction:

```
def buildSideOperatorY(nxp,nyp):
  nxc=nxp-1
  outcome=np.zeros((nxc*nyp,nxp*nyp),float)
  outcome[0:nxc*nyp,0:nxc*nyp] += 0.5 * np.kron( np.identity(nxc) ,
  ↪  np.identity(nyp) )
  outcome[0:nxc*nyp,nyp:nxp*nyp] += 0.5 * np.kron( np.identity(nxc) ,
  ↪  np.identity(nyp) )
  return (outcome)
```

Here the strategy is similar, but the blocks are made of shifted identity matrixes. These operators can now be used to calculate the x and y side average, and therefore the cell average of all quantities:

```
sideXOp=buildSideOperatorX(nxp,nyp)
vxL=vx.reshape(nxp*nyp)
vxSxL=np.dot(sideXOp,vxL)
vxSx=vxSxL.reshape(nxp,nyc)

sideYOp=buildSideOperatorY(nxp,nyp)
vyL=vy.reshape(nxp*nyp)
vySyL=np.dot(sideYOp,vyL)
vySy=vySyL.reshape(nxc,nyp)
```

Where L stands for long, in the sense that the 2D data are stretched in a 1D direction. One can immediately verify that the result is identical to the calculations in Sect. 8.1, but now made with a generic operator. This same approach can now be extended to calculating the derivative, with the similar functions:

```
def buildDxOperator(nxp,nyc,dx):
  nxc=nxp-1
  outcome=np.zeros((nxc*nyc,nxp*nyc),float)
  outcome[0:nxc*nyc,0:nxc*nyc] -=
  ↪  1.0/dx*np.kron(np.identity(nxc),np.identity(nyc))
  outcome[0:nxc*nyc,nyc:nxp*nyc] +=
  ↪  1.0/dx*np.kron(np.identity(nxc),np.identity(nyc))
```

```
    return (outcome)

def buildDyOperator(nxc,nyp,dy):
    nyc=nyp-1
    block=np.zeros((nyc,nyp),float);
    block[0:nyc,0:nyc]-=1.0/dy*np.diag(np.ones(nyc),0)
    block[0:nyc,1:nyp]+=1.0/dy*np.diag(np.ones(nyc),0)
    return (np.kron(np.identity(nxc),block))
```

Where the only difference with the above calculations are the constants that multiply the arrays and the number of cells in the y and x direction, respectively. These four operators allow us to recalculate strain rates, this time just by multiplying the operators by the velocity field, obtaining the same result as the direct approach:

```
# calculate the xx strain rate
DxOp = buildDxOperator(nxp,nyc,dx)
DvxDxL = np.dot(DxOp,vxSxL)
DvxDx = DvxDxL.reshape(nxc,nyc)

# calculate the yy strain rate
DyOp = buildDyOperator(nxc,nyp,dy)
DvyDyL = np.dot(DyOp,vySyL)
DvyDy = DvyDyL.reshape(nxc,nyc)
```

How to calculate the xy Strain Rate is shown in Sect. 8.2.1.

What is the goal of creating these operators? The reason is that these linear operators can be flexibly combined together in order to solve large problems and solve them as inverse problems, split the operations on many processors, insert material properties like variable viscosity or hydraulic conductivity, as I will show in the next chapters.

For example, we can invert our matrix-operator and find the velocity field, if enough data are available. Normally we do not know the strain rate, but we know the forcing, and therefore we can relate it to the stress and invert these equations to find the velocity field. To show another example, here I create two *cell-centered derivative operators* by combining the above created operators and use them to calculate the xy strain rate of the above flow, and verify that it is correct:

```
cellDxOp = np.dot(DxOp,sideXOp)
cellDyOp = np.dot(DyOp,sideYOp)
print ('error in SRxx: ', np.max(np.reshape(np.dot(cellDxOp,vxL),
    ↪ (nxc,nyc))-DvxDx), np.min(Dvx-DvxDx))
```

where the last line gives on my computer an error of the order of 10^{-15}. This is the approach that we will look at in the next two chapters.

At this point, it is extremely important that this book is not used as a *reading company* but as a manual, which means that the student is expected to repeat the above examples and to solve the problems at the end of the book, in order to gain a clear understanding of these procedures. It will be soon clear that the use of operator is central in the rest of the book. One can also clearly understand that besides being simple and compacts, operators can be easily parallelized (although we will see in the next section that we need to look first at their *sparse* description). For all these reasons, the more one can invest time here in testing and playing with these toys, the better will the more advanced (nonlinear flow) topics will appear.

8.2.1 Sparse Derivative Operator

Until now, we did not spend much time looking at how efficient operators are. In fact, although they are a very elegant and efficient way to operate on our data, the ones that we have built are far from being computationally optimal. We can calculate the time required to perform a full calculation of xy strain rates using the above described operators:

```
# calculate the xy strain rate
vxL=vx.reshape(nxp*nyp)
vyL=vy.reshape(nxp*nyp)
vySxL = np.dot(sideXOp,vyL)
vxSyL = np.dot(sideYOp,vxL)
DvyDxL = np.dot(DxOp,vySxL)
DvxDyL = np.dot(DyOp,vxSyL)
SRxyL = 0.5*(DvyDxL+DvxDyL)
SRxy = SRxyL.reshape(nxc,nyc)
```

On my laptop, these operations require about 72 ms with 50×100 cells, although the direct calculation took half of a millisecond with 200×200 cells. The reason of this huge difference is that we are creating giant operators of size $nx \times ny$. To reduce this time we have to start using *sparse matrices*. Let us see how we can build the same two operators of Problem 8.3 that allow us to do any derivative operation, just using sparse operators. Of the many sparse arrays that exist in Python, the CSR format (Compressed Sparse Column format) is specially suitable for fast matrix vector products. Since all our matrix come from some combinations of diagonal matrices, we can combine them into a CSR format. Let us see how to do it.

The strategy is simply to proceed in blocks. I show here how to make them with one technique based on diagonal sparse matrices, but other approaches (for example, building directly a CSR sparse matrix) can be as efficient as this one, or even more, and the reader is welcome to test them and find better implementations. Following the approach described earlier, X axis operator will be based on large diagonals, while the Y axis operator on *kron()* of blocks that will be made of sparse diagonal matrices. Here, there is a possible formulation:

```
from scipy.sparse import sparse

def buildSparseOperatorXaxis(nx,ny,k1,k2):
  nx1=nx-1
  firstDiag=np.ones(nx*ny,float)*k1
```

```
  secondDiag=np.ones(nx*ny,float)*k2
  offsets=np.array([0,ny])
  return ( sparse.dia_matrix(([firstDiag,secondDiag],offsets),
  ↪  shape=(nx1*ny,nx*ny)).tocsr() )

def buildSparseOperatorYaxis(nx,ny,k1,k2):
  ny1=ny-1
  firstDiag=np.ones(ny,float)*k1
  secondDiag=np.ones(ny,float)*k2
  offsets=np.array([0,1])
  block = sparse.dia_matrix(([firstDiag,secondDiag],offsets),
  ↪  shape=(ny1,ny)).tocsr()
  return (sparse.kron(sparse.eye(nx),block))
```

One can notice several aspects here. One is that *firstDiag* and *secondDiag* are longer than what they should be. The reason is that the command *offset* shifts the diagonal up, not to the right as we would like, so we have to create a longer *secondDiag* in order to cover the entire second diagonal. At the same time both diagonal arrays have to be long equal, therefore both are long and then cut when the matrix is created.

Using these functions, it is possible to calculate the operator for the xy Strain Rate in the following form:

```
vxL=vx.reshape(nxp*nyp)
vyL=vy.reshape(nxp*nyp)

DxOp = buildSparseOperatorXaxis(nxp,nyc,-1.0/dx,1.0/dx)
DyOp = buildSparseOperatorYaxis(nxc,nyp,-1.0/dy,1.0/dy)
sideXOp =buildSparseOperatorYaxis(nxp,nyp, 0.5, 0.5)
sideYOp = buildSparseOperatorXaxis(nxp,nyp, 0.5, 0.5)

cellDxOp = DxOp.dot(sideXOp)
cellDyOp = DyOp.dot(sideYOp)

SRxy = np.reshape(0.5*cellDyOp.dot(vxL)+0.5*cellDxOp.dot(vyL), (nxc,nyc))
```

Which on my laptop requires only 3 ms for 100×50 cells, whereas the same approach using dense operators would require about 16 seconds, due to the extremely demanding matrix–matrix multiplication! We are now ready to head towards solving problems for which we know physical properties and forces and we want to find the flow.

8.3 Reversible and Irreversible

Not every deformation is identical. *Reversible* deformation, like the one of rubber, is called *elastic*, and is generally described as a function of strain. Strain holds an information of how far we have deformed our media from the initial undeformed structure, and therefore of how much energy we need to restore it. *Irreversible* deformation, like the one of flowing corn sirup, is called *viscous* and is described only as a function of strain rate.

Reversible deformation is associated to a *potential* that quantifies how much energy is stored in the deformed material. For example, the incredible power of Earthquakes are due to the sudden release of the elastic energy stored in the deformation of the lithosphere. *Irreversible* deformation is instead associated to *dissipative* energy and therefore quantifies how much energy will become heat. In fact, the equation associated to the evolution of temperature is typically called the *Energy equation*.

It is very important for us that the strain, strain rate are different from one point to the other of a body, implying that stores and dissipated energies should also better represented as *local* quantities, i.e., associated to a small parcel of solid or fluid. This is done by defining *energy densities*, either *stored*, e.g., $\frac{1}{2}\rho\varepsilon^2$ in linear elasticity, or *dissipated*, e.g., $\frac{1}{2}\eta\dot{\varepsilon}^2$ for a linearly viscous system.

Due to thermodynamic constrains, irreversible systems will tend maximize the dissipated energy. It is, therefore, important to calculate strain and strain rates correctly and precisely in order to monitor these quantities. The rest of this chapter will look at a new more compact and very efficient way to calculate these key quantities in continuous mechanics, on a cartesian mesh.

In many geodynamic systems elastic strains are very small and the linear elastic approximation is sufficient (with the notable exception of volumetric strains in the Earth deep interiors that can be huge). Instead dissipative systems involve very large strains and require a general nonlinear description. We will see how many systems can be initially solved as linear, while the nonlinear contribution can be added as a *perturbation*, and the final solution obtained by iterating the process until *convergence* is achieved.

Summary

- Flow laws can be reversible or irreversible. Reversibility can be used to check the quality of a simulation. Generally linear Stokes flow is reversible, while nonlinear Stokes Flow is not reversible.
- All differential operations on a function in any number of dimensions can be written as an Operator, which has always a matrix form.
- Operators form a group, therefore they can be applied one to the other to form complex operators such as Divergence, Gradient, and Laplacian.
- Applied to flow laws, operators allow immediately to calculate Strain Rates, deviatoric, deformation, volumetric variations, and more.
- Operators describing Derivative Tensors can be Sparse or Dense, and for large problems to use the sparse ones is mandatory.

The routines introduced in this and the next chapter will be reused in the rest of the book, therefore it is important to create, optimize and store a library that can be accessed to accelerate the development of more sophisticated codes.

Problems

8.1 Verify that the calculation of the continuity equation with a pure rotation is correct also neglecting the cell-centered calculation. This is due to the simplicity of the flow.

8.2 Create a more complex flow to verify the validity of the cell-centered approximation. Remember that its divergence has to be zero. A comfortable Python-based software to calculate and test analytical functions is SAGE (`http://www.sagemath.org/`), similar to *Mathematica*.

8.3 The four functions *buildSideOperatorX buildSideOperatorY buildDxOperator* and *buildDyOperator* have a similar structures. Can you write two functions that, when called, will create one or the other operator? Is it possible to have only one function that does it?)

8.4 How would you combine the two operators *cellDxOp* and *cellDxOp* to create a unique large operator that creates an array of xx, yy, and xy strain rates from a $nx \times ny$ array of velocities?

8.5 As an exercise, take the sparse and dense formulations for calculating one of the strain rates and test how the calculation time scales with growing nx and ny. Theoretically, it should scale quadratically with the dense formulation and linearly with the sparse one.

Chapter 9
Laplacian Operator and Diffusion

Abstract Diffusion is the most common process in Geodynamics. Examples of explicit and implicit diffusion solvers are shown. Tests show that explicit solvers suffer of stability issues, while implicit ones are generally conditionally stable. The second part of the chapter shows through several examples how to create sparse matrix operators that minimized memory occupation and computing time. Finally it is shown how the Stokes Flow can be solved by solving for the stream-function, using a double application of the diffusion matrix operator, called biharmonic operator.

The most important physical process in Geodynamics is probably the second law of thermodynamics that says that the disorder in the universe constantly grows. In fact physics tells us that the second law is the main indication of what is time, given that the growth of entropy is the only nonreversible process among the fundamental laws of physics.

Among the geodynamic processes, the growth of the entropy manifests itself mainly by diffusional processes. Diffusion is in fact a throughly irreversible processes, characterized by the chaotic mix of particles of different nature, like the diffusion of a drop of color in water, or by the flow of heat from a hot to a cold body, or by the diffusion of a liquid in a porous media (Darcy's flow). All these phenomena happen in any long-term process, making diffusion an ubiquitous actor in geodynamics.

Historically, the similarity between the diffusion of a fluid into another fluid and the diffusion of heat in a solid convinced scientists that heat was nothing else than a invisible fluid, called *calorico* that was transporting this energy, heat. Maxwell and Boltzmann were the pioneers who figured out first the equivalence between heat and kinetic energy, and that therefore the temperature was nothing else than the measure of the kinetic energy of every single mode of motion of particles. For example, for a monoatomic Noble gas the average kinetic $\frac{1}{2}/\mathrm{mv}^2$ of each atom is $\frac{3}{2}kT$ where k is the Boltzmann constant ($1.38 \cdot 10^{23}$ J/K) and m is the mass of the atom.

G. Morra, *Pythonic Geodynamics*, Lecture Notes in Earth System Sciences,
DOI 10.1007/978-3-319-55682-6_9

9.1 Diffusion Processes in Geodynamics

In geodynamics, we mostly deal with crystalline rocks for which the heat is stored as the kinetic energy of vibrating atoms whose position oscillates around the equilibrium position in the lattice. This kinetic energy is transmitted to the neighboring atoms through interatomic electric forces. Also, here the kinetic energy associated to the atomic oscillation is $\frac{3}{2}kT$. The fact that all these different phenomena are controlled by the same laws is at the same time a truly remarkable aspect of physics and also greatly simplifies the life of geodynamic modelers because it makes the numerical modeling of these effects very similar.

For example, after their formation at mid ocean ridges, crust and lithosphere cool mostly due to diffusion of heat at the surface, while the plates slide horizontally for 100 million years or more. Heat diffusion also balances the convection in the mantle and reduces its intensity. Chemical diffusion in planetary interiors can be very complex because several materials diffuse a different speed, creating nonlinear effects, which ultimately reduce to the same general law. In general every material that we, as humans, will try to constrain to protect the environment (e.g., radioactive, pollution) will ultimately leak out. For example, storing CO_2 in rocks is limited by the efficiency of the diffusion in rock, as well as the storage of geothermal heat at the surface. We cannot stop these processes, but we can monitor and limit them.

Another type of phenomena that surprisingly follows the diffusion law is random walk. It was Einstein, who first understood that the seemingly chaotic motion of a large particle immersed in a fluid, also called Brownian motion, can be described by their being hit by vibrating molecules. By calculating the overall effect of these collisions he devised a way to calculate the effective diffusion of these particles and the effective viscosity of the fluid, only from the Avogadro number N of particles in a mole and from the size of the molecules. This calculation concluded his PhD thesis, demonstrated the existence of atoms, showed how their large scale effect is a diffusional process and was the discovery for which he was awarded the Nobel price in physics in 1921 [91].

Similarly to random walk, the diffusion of the single parcels of fluid in porous rocks follows complex pattern that overall average to a diffusion process. Hydrogeological flow and geothermal flow are driven diffusion processes that acquire a direction because of the gravity forcing. In the last part of the book, I will discuss the microscopic aspects of this multicomponent and multiphase flow and their modeling.

Since all forms of diffusion follow the same law, I will discuss here how to solve the heat equation, without loosing generality. This can be generalized to any other diffusion process. Regardless on which phenomena it refers to, ultimately diffusion is described by a *coefficient of diffusion* that can depend on any macroscopic variable. For heat diffusion with thermal conductivity k, density ρ and heat capacity at constant pressure C_p the thermal diffusivity is defined as $\kappa = k/(\rho C_p)$. Let us see how to prove this.

If diffusion is the only thermal process, mathematically the evolution of temperature in a cell depends only on the difference between the heat that is coming and the one that is leaving the cell. In one dimension this can be simply expressed by calculating the heat flux q as

$$q_x = -k\frac{\partial T}{\partial x} \tag{9.1}$$

where k is the thermal conductivity and the mines indicates that the heat goes from the region of higher to lower temperatures. The temperature change in cell of material that is sufficiently small that we can consider temperature gradient constant within the cell, is

$$C_p\rho\delta T = \frac{\partial q_x}{\partial x}\delta t \tag{9.2}$$

where the derivative of q_x indicates the differential heat flow. By dividing by δt one obtains

$$C_p\rho\frac{\partial T}{\partial t} = -\frac{\partial}{\partial x}\left(k\frac{\partial T}{\partial x}\right) \tag{9.3}$$

where in the left side term the temperature $T = T(x, t)$ that is both a function of space and time, and it is derived *partially* by time, where the partial derivatives means that the cell does not move. This way to average quantities is called Eulerian. It is the point of view of an observer that does not move with the fluid.

If C_p and ρ are constant we can express this equation in function of a unique parameter and naturally extended them in 2D and 3D:

$$\begin{aligned}
\frac{\partial T(x,t)}{\partial t} &= \frac{\partial}{\partial x}\left(k\frac{\partial T}{\partial x}\right) \\
\frac{\partial T(x,y,t)}{\partial t} &= \frac{\partial}{\partial x}\left(k\frac{\partial T}{\partial x}\right) + \frac{\partial}{\partial y}\left(k\frac{\partial T}{\partial y}\right) \\
\frac{\partial T(x,y,z,t)}{\partial t} &= \frac{\partial}{\partial x}\left(k\frac{\partial T}{\partial x}\right) + \frac{\partial}{\partial y}\left(k\frac{\partial T}{\partial y}\right) + \frac{\partial}{\partial z}\left(k\frac{\partial T}{\partial z}\right)
\end{aligned} \tag{9.4}$$

Here x, y, z, and t indicate the *independent* variables. One of the most important characteristics of these equations are the presence of the diffusivity inside the parenthesis. If the diffusivity is constant respect to space, we could extract it from the derivative and these equation would immediately simplify into:

$$\begin{aligned}
\frac{\partial T(x,t)}{\partial t} &= k\frac{\partial^2 T}{\partial x^2} \\
\frac{\partial T(x,y,t)}{\partial t} &= k\left(\frac{\partial^2 T}{\partial x^2} + \frac{\partial^2 T}{\partial x^2}\right) \\
\frac{\partial T(x,y,z,t)}{\partial t} &= k\left(\frac{\partial^2 T}{\partial x^2} + \frac{\partial^2 T}{\partial x^2} + \frac{\partial^2 T}{\partial x^2}\right)
\end{aligned} \tag{9.5}$$

However the most general case, and also the most interesting from the geodynamic point of view, is the one in which k is nonhomogenous, and possibly nonlinear depending on some physical parameters.

We can already envisage here how the operators introduced earlier can be applied here. If the diffusivity is in the parenthesis we can first create a derivative operator, then apply the diffusivity, then apply the derivative operator again and so obtain the temperature increment in time. This approach, called *explicit*, is however effective only for very small time steps, as we will see in Sect. 9.2.

We have considered above only the flow of heat. In general both heat and material can flow, and the material might transport heat with it, creating what is called *advection* of heat. This case is a simple extension of the one above, where we can replace the *partial* derivative in the terms on the left with a *full* derivative, using the standard definition of derivatives. In 1D, for example, it looks like

$$\frac{dT(x,t)}{dt} = \frac{\partial T}{\partial t} + \frac{\partial T}{\partial x}\frac{\partial x}{\partial t} = \frac{\partial T}{\partial t} + v_x \frac{\partial T}{\partial x} \tag{9.6}$$

where v_x is the x component of the velocity vector. The extension in 2D and 3D is obvious. This way to calculate a derivative is called *Lagrangian*, in opposition to the *Eulerian* viewpoint introduced earlier. In practice, the Eulerian viewpoint means that I stay on the side of a river and look at water flowing and bringing everything with it. The Lagrangian viewpoint is equivalent to be on a canoe without a paddle and letting the stream taking me with it. By combining Eqs. (9.6) and (9.4) and taking the velocity bearing terms to the right hand side one obtains:

$$\begin{aligned}
\frac{\partial T(x,t)}{\partial t} &= \frac{\partial}{\partial x}\left(k\frac{\partial T}{\partial x}\right) - v_x\frac{\partial T}{\partial x}\\
\frac{\partial T(x,y,t)}{\partial t} &= \frac{\partial}{\partial x}\left(k\frac{\partial T}{\partial x}\right) + \frac{\partial}{\partial y}\left(k\frac{\partial T}{\partial y}\right) - v_x\frac{\partial T}{\partial x} - v_y\frac{\partial T}{\partial y}\\
\frac{\partial T(x,y,z,t)}{\partial t} &= \frac{\partial}{\partial x}\left(k\frac{\partial T}{\partial x}\right) + \frac{\partial}{\partial y}\left(k\frac{\partial T}{\partial y}\right) + \frac{\partial}{\partial z}\left(k\frac{\partial T}{\partial z}\right) - v_x\frac{\partial T}{\partial x} - v_y\frac{\partial T}{\partial y} - v_z\frac{\partial T}{\partial z}
\end{aligned}$$

Again we can envisage, just by looking at this formulation, how we can use the operator matrix formulation to calculate the right-hand side, comprised of the advection terms, and then use it to calculate the Temperature increment. Although this technique is intrinsically unstable, its implementation is a very interesting and useful exercise, and are a stimulus to learn the more challenging implicit techniques.

9.2 Explicit Diffusion Implementation

Explicit numerical formulations are the easiest to implement and when effective, extremely useful. Unfortunately a *stability* issue is common to most of them, and it is important to understand from the beginning what is the cause.

To familiarize with the problem of stability it is easier to look at the simplest equation, the 1D version of the non advecting heat flow equation (9.5) with constant coefficients. We can immediately extend the first derivative in time and the second derivative in space as

$$\frac{T^{t+\Delta t}(x_i) - T^t(x_i)}{\Delta t} = k\frac{T^t(x_{i+1}) - 2T^t(x_i) + T^t(x_{i-1})}{\Delta x^2} \tag{9.7}$$

If the space discretization is constant, it is possible to reformulate the temperature at the time $t+\Delta t$ in function of the temperature at the time t and of a unique parameter $r = k\Delta t/\Delta x^2$:

$$T^{t+\Delta t} = T^t(x_i) + r * \left[T^t(x_{i+1}) - 2T^t(x_i) + T^t(x_{i-1})\right] \tag{9.8}$$

Let us now use the techniques that we learned in the past chapters to create a straightforward implementation of this formula:

```
import numpy as np
import matplotlib.pyplot as plt

nxp=101; nxc=nxp-1; xTot=1.0;
dx=xTot/nxc; xMin=-xTot/2.0; xMax=xTot/2.0
steps=1000; totalTime=1.0; plotEverySteps = 100;
deltaTime=totalTime/steps
diffusivity=1e-1; r = deltaTime*diffusivity/(dx*dx)

X = np.arange(nxp)*dx+xMin
T = np.sin(2*np.pi*X)

DT=np.zeros(nxp)
for time in np.arange(steps):
  DT[1:nxp-1]=T[0:nxp-2]-2*T[1:nxp-1]+T[2:nxp]
  T += r * DT
```

where X is space, T is the temperature, and the system is initialized with a smooth sinusoidal initial temperature. The system is explicit in the sense that the Laplacian of the temperature field, used to calculated the Temperature increment, is obtained at the old time step and then used to calculate the new temperature field.

By playing with any parameter that changes the coefficient r, one discovers that regardless of the quality of the result as soon as the coefficient r is above the value 0.5, the system becomes completely unstable. An example created with the above routine is shown in Fig. 9.1.

This is a typical example of counterintuitive aspects of Numerical modeling. Why does this instability appear? What happens is that for $r > 0.5$ small perturbations amplify exponentially. This can be understood from the line $T += r * DT$. Given that DT for certain values of the function T becomes two times the difference between the temperature in one point and its neighboring one, when r is greater than 0.5 then $r * DT$ becomes greater than this difference and therefore at the next time step the

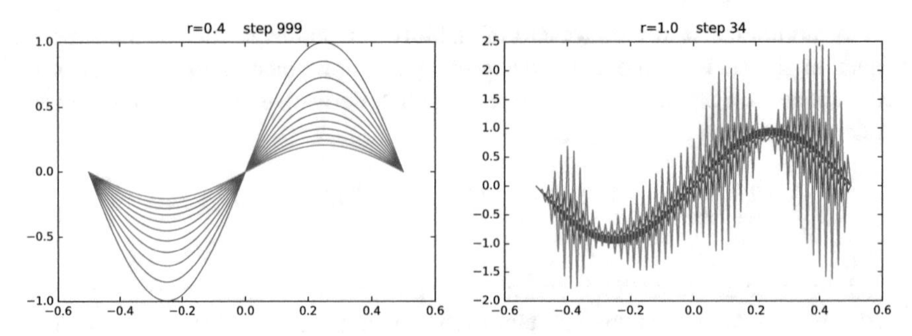

Fig. 9.1 *Left* 1000 timesteps of a stable diffusion simulation with a sinusoidal initial condition. Stability is guaranteed by the value of $r = 0.4$ that is less than 0.5. *Right* step 35 of the same simulation, this time with $r = 1.0$. The simulation develops an instability that becomes visible and comparable to the amplitude of T immediately after the 30th step and continues to grow exponentially

difference between the temperature at one node and the temperature of the next node becomes greater, triggering this instability. The exponential and destructive growth of these instabilities is evident in the right image of Fig. 9.1.

9.3 Explicit Formulation Using Operators

Before discussing about how the stability problem can be solved, let us look at how we can use the operators that we introduced in the past chapter to write the 2D version in Python of the explicit diffusion algorithm. I will assume that *numpy* and *matplotlib.pyplot* are loaded and that *buildSparseOperatorXaxis* and *buildSparse-OperatorYaxis* are defined as in Chap. 8. We first set the variables and create the arrays:

```
nxp=51; nxc=nxp-1; xTot=1.0; nyp=51; nyc=nyp-1; yTot=1.0;
dx=xTot/nxc; xMin=-xTot/2.0; xMax=xTot/2.0
dy=yTot/nyc; yMin=-yTot/2.0; yMax=yTot/2.0
steps=1000; totalTime=1.0; plotEverySteps = 100;
deltaTime=totalTime/steps; diffusivity=5e-2;
r = deltaTime*diffusivity/(dx*dy)

X=np.arange(nxp)*dx+xMin
Y=np.arange(nyp)*dy+yMin
T=np.outer(np.sin(2*np.pi*X),np.sin(2*np.pi*Y))
```

where the temperature is defined by a combination of sinusoidal waves. It is possible of course to use any other initial condition.

To create the necessary operators, we need to combine derivative operators. As we did for the Strain Rate calculation in Sect. 8.3.1 here we have to build the derivative operator in x and y. This time, however, we need to apply both x and y operators twice. Since the first derivative has to be centered in the cell, it will have a smaller size, being $[(nxp - 1) \times (nyp - 1)] \times [(nxp - 1) \times (nyp - 1)]$. These operators are called here $cellDxOp$ and $cellDyOp$.

The second derivative needs to be applied to a smaller set of points (only the cell centers) and will be therefore centered at the internal nodes of the domain. These operators are called here $cellDxOpSecond$ and $cellDyOpSecond$. Finally, the four operators are combined to create the *Laplacian* operator, which corresponds to $\partial^2/\partial x^2$ and $\partial^2/\partial y^2$:

```
# create operator for the first derivative
DxOp = buildSparseOperatorXaxis(nxp,nyc,-1.0/dx,1.0/dx)
DyOp = buildSparseOperatorYaxis(nxc,nyp,-1.0/dy,1.0/dy)
sideXOp =buildSparseOperatorYaxis(nxp,nyp, 0.5, 0.5)
sideYOp = buildSparseOperatorXaxis(nxp,nyp, 0.5, 0.5)

cellDxOp = DxOp.dot(sideXOp)
cellDyOp = DyOp.dot(sideYOp)

# create a smaller operator for the second derivative
DxOpSecond = buildSparseOperatorXaxis(nxp-1,nyc-1,-1.0/dx,1.0/dx)
DyOpSecond = buildSparseOperatorYaxis(nxc-1,nyp-1,-1.0/dy,1.0/dy)
sideXOpSecond =buildSparseOperatorYaxis(nxp-1,nyp-1, 0.5, 0.5)
sideYOpSecond = buildSparseOperatorXaxis(nxp-1,nyp-1, 0.5, 0.5)

cellDxOpSecond = DxOpSecond.dot(sideXOpSecond)
cellDyOpSecond = DyOpSecond.dot(sideYOpSecond)

# apply operators together to create x and y second derivative
LaplacianOp = cellDxOpSecond.dot(cellDxOp) + cellDyOpSecond.dot(cellDyOp)
```

This example is clear and straightforward, however there is an even more efficient and elegant way to create a Laplacian operator. Laplacian are in fact also the inner product of two gradients. That more compact version of the same calculation is the following, which gives the same result:

```
# Functional approach to the calculation of the Laplacian
def sparseGradientOperator(nxp,nyp,dx,dy):
  DxOp = buildSparseOperatorXaxis(nxp,nyp-1,-1.0/dx,1.0/dx)
  DyOp = buildSparseOperatorYaxis(nxp-1,nyp,-1.0/dy,1.0/dy)
  sideXOp=buildSparseOperatorYaxis(nxp,nyp, 0.5, 0.5)
  sideYOp=buildSparseOperatorXaxis(nxp,nyp, 0.5, 0.5)
  return(DxOp.dot(sideXOp),DyOp.dot(sideYOp))

def sparseLaplacianOperator(nxp,nyp,dx,dy):
  (cellDxOp,cellDyOp)=sparseGradientOperator(nxp,nyp,dx,dy)
  (cellDxOpSmall,cellDyOpSmall)= sparseGradientOperator(nxp-1,nyp-1,dx,dy)
  return(cellDxOpSmall.dot(cellDxOp)+cellDyOpSmall.dot(cellDyOp))
```

Once the Laplacian operator is created, the explicit calculation of the temperature increment proceeds as before. It is possible to verify as well the role of the critical r parameter.

```
for time in np.arange(steps):
  T_Laplacian = LaplacianOp.dot(T.reshape(nxp*nyp))
  T[1:nxp-1,1:nyp-1] += diffusivity*deltaTime *
  ↪  T_Laplacian.reshape(nxp-2,nyp-2)
```

where the Laplacian only updates the internal values of the nodes of the Temperature. In a real scenario, the external temperature nodes could be set to values that depend on Boundary conditions. This sophisticated approach does not solve stability, but these operators will be very useful for developing implicit formulations. The reader is invited to test this formulation to empirically find the critical r.

It is clear for this approach that since instabilities arise for very small time steps, it would be desirable to use another incremental procedure for the diffusion equation. However, it is important to keep in mind that explicit solvers are really extremely fast and for small time steps produce accurate solutions, therefore when other complications in the numerical model such as severe nonlinearities in the solution of the momentum equation limit time steps below critical stability, the use of explicitly temperature diffusion is still commonly used. Clearly in this case it is essential to check at every time step that the critical Δt is never reached.

9.4 Implicit Formulation

The only way to overcome the limitations of the explicit formulation is to embed the future solution into the calculation of the temperature (or other field) diffusion increment. This approach requires in some way to invert the *Laplacian* operator that we introduced in Sect. 9.2. This more computationally demanding approach has, however, the ability to offer more accurate solutions and in particular it is unconditionally stable for any Δt.

If we rewrite the 1D discretization of the diffusion equation (9.8), changing the right-hand side and calculating it on the final time step instead of the initial one the equation appears as follows:

$$\frac{T^{t+\Delta t}(x_i) - T^t(x_i)}{\Delta t} = k\frac{T^{t+\Delta t}(x_{i+1}) - 2T^{t+\Delta t}(x_i) + T^{t+\Delta t}(x_{i-1})}{\Delta x^2} \tag{9.9}$$

Now, following the same procedure of Sect. 9.2 we rewrite the new values of the field T in function of the old one as well as of the new one:

$$T^{t+\Delta t}(x_i) = T^t(x_i) + r * \left[T^{t+\Delta t}(x_{i+1}) - 2T^{t+\Delta t}(x_i) + T^{t+\Delta t}(x_{i-1})\right] \tag{9.10}$$

where again $r = k\Delta t/\Delta x^2$. Contrary to the explicit formulation, the new values of the field T depend on the old and on the new values themselves. As we see in the following implementation, this inhibits the instabilities that we saw before to arise.

To solve the above equation we need to group all the terms containing $t + \Delta t$ that appear in Eq. (9.10):

$$(1 + 2r)\, T^{t+\Delta t}(x_i) - r T^{t+\Delta t}(x_{i+1}) - r T^{t+\Delta t}(x_{i-1}) = T^t(x_i) \tag{9.11}$$

Since the terms at the next time step are unknown, this equation cannot be written as a simple increment to the old time step, but it requires the solution of N coupled linear equations, representing the new temperature vector as an evolution of the old temperature field. This has an operator formulation.

Calling $\mathbf{T}^t$ the vector of all solutions at the time t, the above equation can be written in a compact form for every point as:

$$(\mathsf{I} + \mathsf{A})\mathbf{T}^{t+\Delta t} = \mathbf{T}^t \tag{9.12}$$

where I is the identity matrix, A represents the evolutionary tridiagonal matrix characterized by $2r$ on the diagonal and $-r$ on the sides of the diagonal, while all the other components of A are zero.

The above system can be solved just by finding the inverse matrix $\mathsf{B} = (\mathsf{I} + \mathsf{A})^{-1}$. A matrix is always invertible if the determinant is nonzero (because the determinant is equal to the product of all the eigenvalues, and if no-eigenvalue is zero it can be inverted), which is our case for every nonzero value of r. This implies that there is a stable solution for every time step, although a larger time step will be less correct. We have so a straightforward expression to use in Python to find the next temperature distribution, given the old one:

$$\mathbf{T}^{t+\Delta t} = (\mathsf{I} + \mathsf{A})^{-1}\mathbf{T}^t = \mathsf{B}\mathbf{T}^t \tag{9.13}$$

For simplification, I will create a function that generates the tridiagonal matrix that I have introduced above by using the feature of the *diag* function in NumPy and in sparse, for comparison. One might wonder why not creating immediately only a sparse matrix and work only with its inverse. The problem is that, as we will see, the inverse of a sparse matrix is not always sparse and if a matrix is dense it is more efficient to treat it as a normal NumPy array. Let's start with a dense matrix:

```
import numpy as np

def tridiagDense(r, nxp, k1=-1, k2=0, k3=1):
  a = np.ones(nxp-1)*(-r)
  b = np.ones(nxp)*(2.*r)
  return np.diag(a, k1) + np.diag(b, k2) + np.diag(a, k3)

nxp=101; nxc=nxp-1; xTot=1.0; dx=xTot/nxc; xMin=-xTot/2.0; xMax=xTot/2.0
steps=10; totalTime=1.0; plotEverySteps = 1; deltaTime=totalTime/steps
diffusivity=1e-1; r = deltaTime*diffusivity/(dx*dx); print('r:',r)

IA = np.identity(nxp)+tridiagDense(r,nxp) # I+A
B = np.linalg.inv(IA) # inverse dense matrix
```

```
# initialization
X = np.arange(nxp)*dx+xMin
T = np.sin(2*np.pi*X)

for time in np.arange(steps):
  T = B.dot(T)  #temperature at t + deltaTime
```

This approach returns immediately the same correct solution of the explicit approach, but now we are not limited anymore to small values of r. One can in fact obtain the same solution that required 1000 steps with the explicit approach, in only 10 steps with the implicit approach. This implies that not only is the implicit approach stable, but it is also very accurate.

The reason for the extraordinary accuracy is largely due to the structure of the inverse matrix. This is not a sparse matrix like the explicit one. The lack of sparsity increases with r, and therefore with its ability to propagate the present solution in the future. This is clearly visible in Fig. 9.2, where (*plt.imshow(B)*) and its central section (*plt.plot(B[npc%2])*) are visualized.

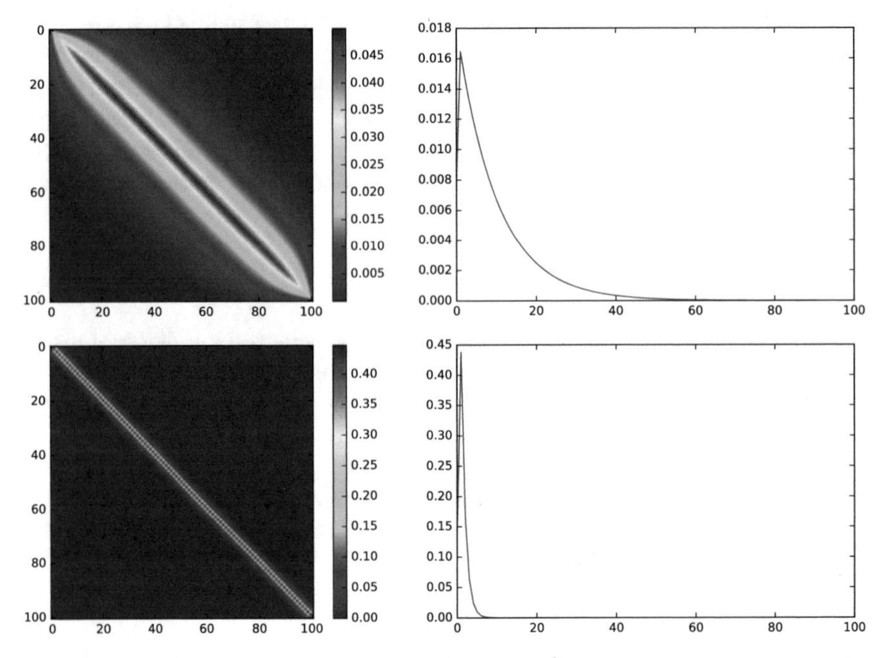

Fig. 9.2 *Left* 2D representation of the matrix $\mathsf{B} = \mathsf{I} + \mathsf{A}^{-1}$ for $r = 100.0$ (*top*) and $r = 1.0$ (*bottom*). One observes that for an implicit algorithm with increasing r (and therefore being far beyond the limiting explicit case), the terms far from the diagonal also increase and the sparse structure is completely lost. *Right* central section of the B matrix where it is visible its decay from the diagonal, smoother with increasing r. For large r matrix sparsity for the inverse is not an advantage anymore

The lack of sparsity of the inverse matrix is important for large problems. The size of the matrix that we have to invert scales like the square of the number of points.

So on a normal computer if we have 10000 points we will have a matrix with 100 millions of entries, which is already close to its memory limits. And this is only equivalent to a 100×100 2D problem.

The use of other libraries for calculating the matrix inverse is not helpful either. For example, we could have defined the tridiagonal matrix as a sparse one with the function *tridiagSparse*, then inverting it with the sparse libraries of scipy. This would have been written, for example, as follows:

```
import scipy.sparse as sparse
import scipy.sparse.linalg as la

def tridiagSparse(r, nxp, k1=-1, k2=0, k3=1):
  a = np.ones(nxp)*(-r)
  b = np.ones(nxp)*(2.*r)
  return sparse.dia_matrix(([a,b,a],[k1,k2,k3]),shape=(nxp,nxp)).tocsc()

IA = sparse.eye(nxp).tocsc()+tridiagSparse(r,nxp)
B = la.inv(IA)
```

However, this does not bring any substantial advantage. A test of the two approaches shows that the sparse matrix approach is about twice as slow as the dense one for $nxp = 1000$, regardless to the value of r.

The generally agreed solution for solving implicitly a system where the inverse matrix is known to be dense, is to not calculate the dense matrix at all. The inverse problem, in fact, can be solved using a sparse solver. In this specific case, instead of calculating the inverse of $I + A$, the inverse problem $(\mathsf{I} + \mathsf{A}) \cdot \mathbf{T}^{t+\Delta t} = \mathbf{T}^t$ is solved as such. This can be done as:

```
IA = sparse.eye(nxp).tocsc()+tridiagSparse(r,nxp)

for thisStep in np.arange(steps):
  T = la.spsolve(IA,T)
```

This approach requires the same computing time of the dense approach for $nxp = 1000$ but it then works also efficiently for greater problem sizes, while the dense matrix approach simply cannot be applied because the matrices reach computer size. A scaling for the three approaches is shown in Fig. 9.3.

9.5 Two-Dimensional Diffusion Equation

By using the same best scaling technique for the one-dimensional diffusion equation (see Fig. 9.3), we can now extend it to the two-dimensional case. Here the matrix that multiplies the $\mathbf{T}$ vector will be the Laplacian one that we calculated in Sect. 9.2. By defining now a function to calculate it, we can find the solution for the implicit problem of an initially Gaussian bell centered temperature distribution around $(0, 0.25)$.

Explicitly we find the 2D equivalent of Eqs. (9.9) and (9.10) as follows:

$$\frac{T^{t+\Delta t}(x_i,y_i)-T^t(x_i,y_i)}{\Delta t}= \tag{9.14}$$
$$k\frac{T^{t+\Delta t}(x_{i+1},y_i)-2T^{t+\Delta t}(x_i,y_i)+T^{t+\Delta t}(x_{i-1},y_i)}{\Delta x^2}+$$
$$k\frac{T^{t+\Delta t}(x_i,y_{i+1})-2T^{t+\Delta t}(x_i,y_i)+T^{t+\Delta t}(x_i,y_{i-1})}{\Delta y^2}$$

and by parametrizing everything with r, this reduces to

$$\begin{aligned}(1+4r)\,T^{t+\Delta t}\left(x_i,y_j\right)-rT^{t+\Delta t}\left(x_i,y_{j+1}\right)-rT^{t+\Delta t}\left(x_i,y_{j-1}\right)\\ -rT^{t+\Delta t}\left(x_{i+1},y_j\right)-rT^{t+\Delta t}\left(x_{i+1},y_j\right)=T^t\left(x_i,y_j\right)\end{aligned} \tag{9.15}$$

The above discretization of the diffusion matrix can be expressed as a matrix operator. Alternatively we could use the operator by applying twice the cell-centered differential operator. The two give very similar, but not identical results.

The difference between the two is illustrated in Fig. 9.4. The approximation in Eq. (9.15) is standard, while the double application of the cell-centered derivative operator gives a *skewed* approximation. The advantage of the skewed approximation is that the first derivative is automatically calculated at the center of the cell, therefore,

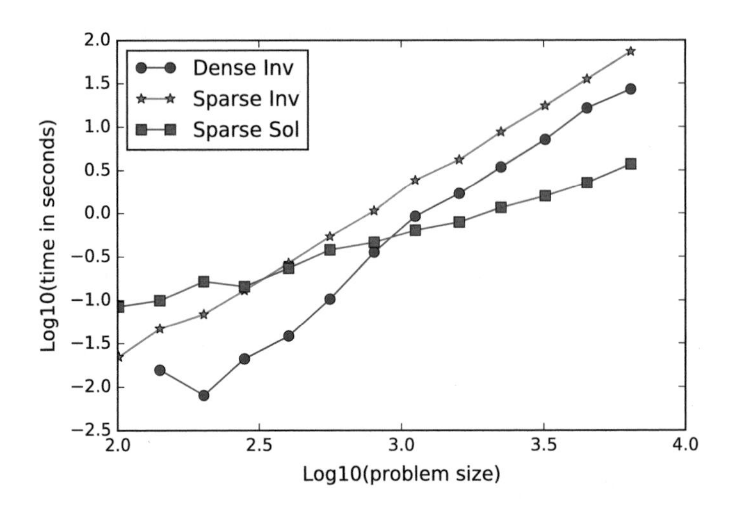

Fig. 9.3 Plot of the Log10 of the Computing time (in seconds) versus log10(nxp) for 1000 steps by using three different strategies for the implicit solution of the diffusion equation. Use a dense matrix is the most efficient only up to about $nxp = 1000$. Above this limit to use the most efficient approach is to solve the equation using the sparse matrix tools. To invert the tridiagonal matrix in sparse format is never efficient because the inverse matrix is always dense. Log10 implies that a jump of one is equivalent to one order of magnitude. This plot was done using MatPlotLib

it is possible to calculate the effect of nonhomogenous diffusion parameters in a simple way, just by using cell-centered values of diffusivity.

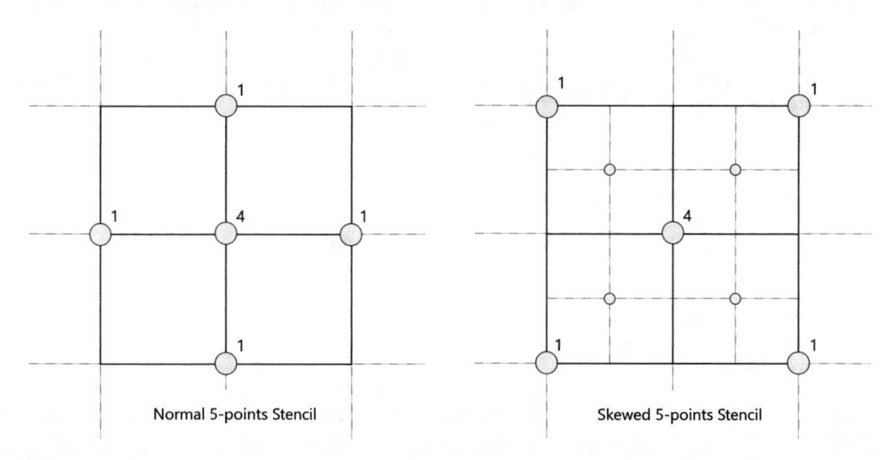

Fig. 9.4 Sketches of the two 5-points stencils used here. The Standard one comes from the expression on Eq. (9.15). The sketch on the *right side* represents the solution emerging from the double application of the cell-centered derivative operator. The first derivative is calculated at the center of the cell, where the *small dots* are, and therefore the second derivative is again at a node. The skewed approach has several advantages

The solution with the two approaches, when diffusivity is constant, is very similar. A very fast implementation of the operator that builds the standard 5-points stencil and uses nonoptimized dense arrays is the following, where I have added extra corrections for the boundaries:

```
def standardLaplacianOperator(nx,ny,r):
  nxy=nx*ny
  L=np.identity(nxy)*4*r

  # four neighbors
  L+=np.diag(np.ones(nxy-1)*(-r),-1)
  L+=np.diag(np.ones(nxy-1)*(-r),+1)
  L+=np.diag(np.ones(nxy-ny)*(-r),-ny)
  L+=np.diag(np.ones(nxy-ny)*(-r),+ny)

  # Side boundaries
  np.fill_diagonal(L[0:ny,0:ny],3.*r)
  np.fill_diagonal(L[nxy-ny:nxy,nxy-ny:nxy],3.*r)

  # Corners
  L[0,0]=2*r
  L[nxy-1,nxy-1]=2*r

  return L
```

This Laplacian Operator can be tested on a standard 50×50 lattice with a high $r = 25.0$, which diffuses very quickly down to a value of the order of 10^{-16} in only 100 steps using the straightforward inversion approach, without showing any instability:

```
from numpy.linalg import inv as denseInv

X=np.arange(nxp)*dx+xMin
Y=np.arange(nyp)*dy+yMin
T=np.outer(np.sin(2*np.pi*X),np.sin(2*np.pi*Y))

L=np.identity(nxp*nyp)+standardLaplacianOperator(nxp,nyp,r)
invL = denseInv(L)

for thisStep in np.arange(steps):
  T = invL.dot(T.reshape(nxp*nyp)).reshape(nxp,nyp)
```

It is a good exercise to try to test other boundary conditions of the *standard LaplacianOperator()* function. One can verify their effectiveness just by visualizing it for a 4×4 or 5×5 lattice. While this approach is simple, it is however limited to systems with constant parameters. To analyze for example convection in the Earth's Mantle or in a magma chamber, or the Darcy flow in a very heterogeneous crust we will need a more flexible tool that can handle large parametric variations.

Let us see how to build these general sparse matrices by using completely vectorized and sparse Side Operators. The first temptation would be to use the *sparseLaplacianOperator* function illustrated in Sect. 9.3. However this operator is defined by a rectangular, not squared, matrix, therefore it cannot be inverted. The reason is that this operator projects a $nxp \times nyp$ array into a $(nxp - 2) \times (nyp - 2)$ array. The missing lines can be added depending on the *Boundary Conditions*. If we assume that the field value are given at the boundary nodes (*Dirichlet* Boundary Conditions), we have only to add empty lines to the Laplacian operator. A simple routine follows that add the *identity* matrix (IM) to the Laplacian operator. In this way the values corresponding to empty lines do not change (they are multiplied by IM):

```
def addBC(operatorWithoutBC,nxp,nyp): # add Boundary Conditions
  nxy=nxp*nyp
  operatorWithBC=sparse.lil_matrix((nxy,nxy))
  indexes = np.arange(nxy)

  # selects only the interior points
  oldLines = (indexes>=nyp) * (indexes<=nxy-nyp) * (indexes % nyp != 0) *
  ↪  (indexes % nyp != nyp-1)
  operatorWithBC[oldLines,:]=operatorWithoutBC[:,:]
  operatorWithBC[~oldLines,~oldLines]=1.0
  return operatorWithBC
```

Observe that I used boolean indexes to find which lines of the new array correspond to the old one, by selecting with four conditions the external nodes of the array.

This method is fully vectorized but not completely efficient. There is still a weakness in the routine because the operation `operatorWithBC[oldLines,:]= operatorWithoutBC[:,:]` between sparse matrices is slowed down by the use of dense matrices to be performed by NumPy. Because of this weakness, it presently works efficiently only up to a resolution of 100×100. This weakness might not remain in future implementation of the *scipy.sparse* library. And the reader is welcome to suggest alternative approaches. For example, it is possible to write a routine that directly adds these lines using the Sparse Matrix notation, based on three arrays, one with the x-index, a second with the y-index and a third with the corresponding nonzero matrix value.

We are not ready to run our sparse 2D solver, similarly to the 1D example. By running this example one can observe how the BC are perfectly dealt:

```
T=np.outer(np.sin(2*np.pi*X),np.sin(2*np.pi*Y))

LaplacianOp=sparseLaplacianOperator(nxp,nyp,dx,dy)
LaplacianOp=addBC(LaplacianOp,nxp,nyp)*r*dx*dx
IA=sparse.eye(nxp*nyp).tocsc()-LaplacianOp  # I-A

for thisStep in np.arange(steps):
  T = la.spsolve(IA,T.reshape(nxp*nyp)).reshape(nxp,nyp)
```

when increasing the computational efficiency for the dense inversion and sparse solver in 2D. The problem size at which the solver starts to be more efficient is much greater. This is due to the fact that the matrix *IA* in 2D is more complex (contains on average 5 terms per row) than in 1D. In 3D, obviously this trend is even greater and to reach a solution with this approach is more challenging.

9.6 Biharmonic Equation

Besides describing the dynamics of diffusion, Laplacian operators can also be applied to the solution of the momentum equation. In fact there is a formulation that uses a *Stream Function* to express two-dimensional flow, in the case of incompressible flow and homogeneous viscosity. The method simply uses the fact that the Laplacian applied to the Laplacian operator, called *biharmonic operator* solves the *Stream Function* equation. Let us see how it works.

The assumption is that there must be a function $\Psi(\mathbf{x}, t)$ whose partial derivatives in space are the velocities:

$$\frac{\partial \Psi}{\partial y} = v_x$$
$$\frac{\partial \Psi}{\partial x} = -v_y$$

This formulation automatically implies that the divergence of the velocity $\frac{\partial v_x}{\partial x} + \frac{\partial v_y}{\partial y} = 0$ is zero, and therefore that the flow is incompressible, regardless to the shape of the function Ψ. We can therefore focus on finding the function Ψ that solves the momentum equation only.

x and y Stokes (momentum) equations, assuming constant viscosity, simplify into:

$$\eta\left(\frac{\partial^2}{\partial x^2} + \frac{\partial^2}{\partial y^2}\right)v_x - \frac{\partial p}{\partial x} = \rho g_x \tag{9.16}$$
$$\eta\left(\frac{\partial^2}{\partial x^2} + \frac{\partial^2}{\partial y^2}\right)v_y - \frac{\partial p}{\partial y} = \rho g_y$$

By adding cross-deriving them, the first respect to y and the second respect to x, the pressure is eliminated and one remains with:

$$\eta\left(\frac{\partial^2}{\partial x^2} + \frac{\partial^2}{\partial y^2}\right)\left(\frac{\partial^2}{\partial x^2} + \frac{\partial^2}{\partial y^2}\right)\Psi = \frac{\partial \rho}{\partial x}g_y - \frac{\partial \rho}{\partial y}g_x \tag{9.17}$$

where we recognize the Laplacian operator applied twice. We can therefore solve this system by creating a new operator, called biharmonic, composed by the Laplacian of the Laplacian. It is also interesting to notice that by applying only one Laplacian one obtains the so-called *vorticity*, that we will analyze more in detail in the last part of the book. The *Stream Function* Ψ allows finally calculating the velocity field. This seemingly twisted approach is in fact very fast and efficient. Let us use the squared *Laplacian* operator to create a solver for the biharmonic equation:

```
def biharmonicOperator(nxp,nyp,dx,dy):
  return (addBC( sparseLaplacianOperator(nxp,nyp,dx,dy),nxp,nyp)**2 )
```

with the operator approach, that was easy and compact. Since we will calculate derivative in the lattice and we may want to project the solution from the cell centers to the mesh, and vice versa. This can be done with two self-explanatory functions:

```
def MeshToVolume(nxp,nyp):
  return buildSparseOperatorYaxis(nxp-1,nyp,0.5,0.5).dot(
  ↪ buildSparseOperatorXaxis(nxp,nyp,0.5,0.5) )

def VolumeToMesh(nxp,nyp):
  return buildSparseOperatorXaxis(nxp,nyp,0.5,0.5).transpose().dot(
  ↪ buildSparseOperatorYaxis(nxp-1,nyp,0.5,0.5).transpose() )
```

In the second function, I have used the *transpose* to project back from the midpoint to the lattice. This creates some information diffusion, that is inevitable since the operator projects less informations to a finer array (from $(nxp - 1) \times (nyp - 1)$ to $nxp \times nyp$.

As an example, let us solve the biharmonic equation for two gravity driven anomalies entrapped in a closed box. This is a common case in geological media.

```
nxp=101; nxc=nxp-1; nxpInt=nxc-1; xTot=1.0;
nyp=101; nyc=nyp-1; nypInt=nyc-1; yTot=1.0;
dx=xTot/nxc; xMin=-xTot/2.0; xMax=xTot/2.0
dy=yTot/nyc; yMin=-yTot/2.0; yMax=yTot/2.0
steps=2; totalTime=2.0; deltaTime=totalTime/steps;

#allocate arrays
X=np.arange(nxp)*dx+xMin; Y=np.arange(nyp)*dy+yMin
vx = np.zeros((nxp,nyp), float)  # x-velocity solution
vy = np.zeros((nxp,nyp), float)  # y-velocity solution
psi = np.zeros((nxp,nyp), float)  # sreamline function

(Dx,Dy)= sparseGradientOperator(nxp,nyp,dx,dy)

# initial conditions, create two density anomalies
density=np.zeros((nxp,nyp),float)
cx1=-0.25;cx2=0.1;cy1=-0.2;cy2=0.3;radius=0.1; #center and radius
XX=np.outer(X,np.ones(nyp));YY=np.outer(np.ones(nxp),Y);
anomalyIndexes = ((XX-cx1)**2+(YY-cy1)**2<radius**2)
anomalyIndexes+= ((XX-cx2)**2+(YY-cy2)**2<radius**2)
density[anomalyIndexes]=1.0

# solve for the streamline function
rhs = VolumeToMesh(nxp,nyp).dot( (1/visc*(gy*Dx-gx*Dy)).dot(
  ↪  density.reshape(nxp*nyp) ))
psi = la.spsolve( biharmonicOperator(nxp,nyp,dx,dy) , rhs )

# extract the velocities
vx = (VolumeToMesh(nxp,nyp).dot(Dy.dot(psi))).reshape(nxp,nyp)
vy = (VolumeToMesh(nxp,nyp).dot(-Dx.dot(psi))).reshape(nxp,nyp)
```

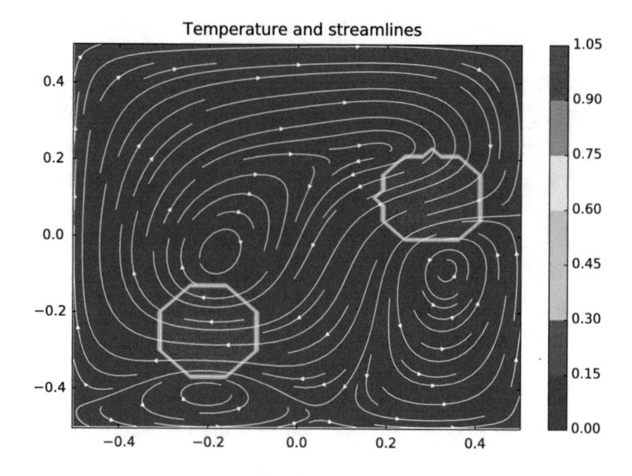

Fig. 9.5 Result of the flow of two particles in a box, using the biharmonic operator. The technique is extremely efficient but limited by the requirement of constant viscosity

The result of this simulation is in Fig. 9.5. The method can be of course combined with the particles method introduced earlier, for example, and obtain a non-diffusive motion of the particles. But we need now to step into a more important problem, that is the case of non-homogeneous momentum equation, i.e., what happens when the strength of the material changes point to point within the domain.

Summary

- Diffusion is the most common phenomenon in Geodynamics.
- Explicit Diffusion solvers are easy to implement but are unstable already for short time increments.
- Implicit Solvers are unconditionally stable and also reliable for very large steps. However their direct application requires calculating dense matrices.
- Iterative solvers, many already present in the SciPy libraries, allow building solvers that scale linearly with the problem size.
- Laplacian Operators can be also defined and built also with derivative operators.
- Using the Biharmonic Operator the same formulation for the Laplacian allows solving the momentum equation for constant viscosity.

Problems

9.1 Consider a squared 2D grid, with constant thermal diffusivity, no advection (velocity), no heat sources, and zero temperature at the boundary. The discretized two-dimensional diffusion equation is the natural extension of Eq. (9.7):

$$\begin{aligned}\frac{T^{t+\Delta t}\left(x_i, y_i\right) - T^t\left(x_i, y_i\right)}{\Delta t} &= k\frac{T^t\left(x_{i+1}, y_i\right) - 2T^t\left(x_i, y_i\right) + T^t\left(x_{i-1}, y_i\right)}{\Delta x^2} + \\ &+ k\frac{T^t\left(x_i, y_{i+1}\right) - 2T^t\left(x_i, y_i\right) + T^t\left(x_i, y_{i-1}\right)}{\Delta y^2}\end{aligned} \tag{9.18}$$

which simplifies as it did the above one assuming $r = k\Delta t/\Delta x^2 = k\Delta t/\Delta x^2$ in case of homogeneous regular discretization ($\Delta x = \Delta y$).

Implement this solution in Python and show that the same instability appears for this system, but for $r > 0.25$. Can you explain why?

9.2 Compare the time required to perform the simulation with a 50×50, 100×100 and 150×150 meshes by using the expression in Eq. (9.18) and the operator approach illustrated in this section. Excluded the time necessary to build the operator, do they scale differently?

9.3 It is possible to create the tridiagonal Laplacian matrix from the 1D operator derivative, applied twice. Can you create these operators?

9.4 It is known that $\frac{1}{1-a} = \sum_{n=0}^{\infty} a^n$. Write a Python program that uses this property to approximate the inverse of $I + A$ with a finite sum of $(-A)^n$ (e.g., up to n = 50) and find the maximum value of r for which this approach is effective. What do you deduce from the result?

Chapter 10
Beyond Linearity

"People take the longest possible paths, digress to numerous dead ends, and make all kinds of mistakes. Then historians come along and write summaries of this messy, nonlinear process and make it appear like a simple, straight line."

—Dean Kamen

Abstract This chapter addresses how to implement a general velocity–pressure formulation non-linear Stokes Flow. It shows how several strategies exist for solving the momentum and the continuity equations simultaneously but that many obstacles are on the path toward a robust solution. Finally a solution is shown with a finite volume fully vectorized implementation that is then applied to classical problems in geodynamics such as mantle convection, particles sedimentation and bubbly flow.

We have until now seen the power of the Laplace operator, and how it captures the phenomenon of diffusion and, using the stream-function approach, the solution of the Stokes equation for the case of homogeneous viscosity. But what happens when the material strength (viscosity, elasticity) or diffusive parameters (thermal or hydraulic conductivity) of the material are not homogeneous?

For example, in the example shown in Sect. 9.6 the spheres could be more viscous than the surrounding fluid (e.g., crystals in magma, stiff in the mantle) or much less viscous (e.g., gas bubble in magma or in water). In this case, the formulation of the stress changes and the derivative of the nonhomogenous and potentially nonlinear parameters enter into the equations to solve. We will see in this chapter that, while linear problems could be approached from similarly many directions, only some well designed algorithms can deal with these nonlinearities.

10.1 Operator Form of the Stokes Equation

Let us go back to the definition of the momentum equation for Stokes Flow, but let's introduce the formulation for Stress when the viscosity is non-homogeneous.

G. Morra, *Pythonic Geodynamics*, Lecture Notes in Earth System Sciences,
DOI 10.1007/978-3-319-55682-6_10

Conceptually Stress is the analogue in continuum mechanics of what the force is in Newtonian mechanics, however in continuum mechanics the forcing term can be in general split in two, a *bulk* term, that is commonly represented as a function of the pressure, and a *deviatoric* term that describes the deviation from the bulk effects. This is particular important at very high pressure as in the deep Earth's interiors where the deviatoric stress is very small compared to the pressure, however, they almost totally control the local flow low. Formally, by calling the total stress tensor τ and the deviatoric one σ the two terms are written as:

$$\tau_{ij} = \sigma_{ij} - p\delta_{ij} \tag{10.1}$$

where the minus conventionally indicates that the compressional pressures are positive.

The momentum equation for a gravity driven system can be written by splitting deviatoric stress and pressure:

$$\frac{\partial \sigma_{ij}}{\partial x_j} - \frac{\partial p}{\partial x_i} = -\rho g_i \tag{10.2}$$

This is the general Stokes equation that we will solve for any constitutive laws that relates our stress with the material deformation. Given that stress is related to deformation (strain rate), it appears from this expression that the most general way to solve Stokes low implies the simultaneous solution of the flow law (given by the local description of the velocities v_x, v_y, v_z) combined with the local solution of the pressure p.

In a viscously deforming material like the Earth's mantle, or the interior of a magma chamber, the relationship between the strain rate $\dot{\varepsilon}$, describing material deformation, and the deviatoric stress σ is defined with the following relationship of proportionality (e.g., [36]):

$$\sigma_{ij} = 2\eta\dot{\varepsilon}_{ij} \tag{10.3}$$

Where η is the viscosity of the material. Since in two-dimensions we want to solve for three unknowns (v_x, v_y and p), the two-dimensional Stokes' equation needs an extra one to close the system. As is standard in Geophysical Fluid dynamics, we will use the incompressible continuity equation. By inserting (10.3) in (10.2), one obtains (in 2D) a 3×3 matrix operator that looks like:

$$\begin{bmatrix} \frac{\partial}{\partial x}\left(2\eta\frac{\partial}{\partial x}\right) + \frac{\partial}{\partial y}\left(\eta\frac{\partial}{\partial y}\right) & \frac{\partial}{\partial y}\left(\eta\frac{\partial}{\partial x}\right) & -\frac{\partial}{\partial x} \\ \frac{\partial}{\partial x}\left(\eta\frac{\partial}{\partial y}\right) & \frac{\partial}{\partial x}\left(\eta\frac{\partial}{\partial x}\right) + \frac{\partial}{\partial y}\left(2\eta\frac{\partial}{\partial y}\right) & -\frac{\partial}{\partial y} \\ \frac{\partial}{\partial x} & \frac{\partial}{\partial y} & 0 \end{bmatrix} \cdot \begin{bmatrix} v_x \\ v_y \\ p \end{bmatrix} = \begin{bmatrix} -\rho g_x \\ -\rho g_y \\ 0 \end{bmatrix} \tag{10.4}$$

where one observes that some *nonlinear* laplacian terms have 2η and others η due to the definition of strain rate. If the viscosity is constant, this reduces to the Eq. (9.17), due to the simplification that $\frac{\partial}{\partial x}\left(\frac{\partial}{\partial y}\right) = \frac{\partial}{\partial y}\left(\frac{\partial}{\partial x}\right)$, combined with the continuity equation. In this case, the operator formulation takes the form:

$$\begin{bmatrix} \eta\frac{\partial}{\partial x}\left(\frac{\partial}{\partial x}\right)+\eta\frac{\partial}{\partial y}\left(\frac{\partial}{\partial y}\right) & 0 & -\frac{\partial}{\partial x} \\ 0 & \eta\frac{\partial}{\partial x}\left(\frac{\partial}{\partial x}\right)+\eta\frac{\partial}{\partial y}\left(\frac{\partial}{\partial y}\right) & -\frac{\partial}{\partial y} \\ \frac{\partial}{\partial x} & \frac{\partial}{\partial y} & 0 \end{bmatrix} \cdot \begin{bmatrix} v_x \\ v_y \\ p \end{bmatrix} = \begin{bmatrix} -\rho g_x \\ -\rho g_y \\ 0 \end{bmatrix} \quad (10.5)$$

This last form is easier to write but not necessary easy to solve, as the next section illustrates.

10.2 Implementation of the Homogeneous Stokes Equation

The solution of this last system should be the same of the streamline approach. Let us test it using our operator approach. Like we built a $nxp \times nyp$ operator in Sect. 9.6 to find the stream-function, here will have to build a $3 * nxp \times 3 * nyp$ operator to simultaneously find velocity and pressure. I will assume the same density anomaly as earlier and only show how to solve the numerical problem and a homogeneous viscosity equal to 1.0.

```
nxy=nxp*nyp
(Dx,Dy) = sparseGradientOperator(nxc,nyc,dx,dy)
#The operators are made square by applying Dirichlet BC
Dx = addBC( Dx.dot( MeshToVolume(nxp,nyp) ),nxp,nyp )
Dy = addBC( Dy.dot( MeshToVolume(nxp,nyp) ),nxp,nyp )

L = addBC( sparseLaplacianOperator(nxp,nyp,dx,dy),nxp,nyp)

fullStokesOperator = sparse.bmat([[L, None, -Dx], [None, L, -Dy],
  ↪ [Dx,Dy,None]]).tocsr()

rhs = np.zeros((3*nxy), float) # right hand side
rhs[0:nxy]= -gx*density.reshape(nxy)
rhs[nxy:2*nxy]= -gy*density.reshape(nxy)

solution = la.spsolve(fullStokesOperator,rhs) # solve

vx=solution[0:nxy].reshape(nxp,nyp)
vy=solution[nxy:2*nxy].reshape(nxp,nyp)
p=solution[2*nxy:3*nxy].reshape(nxp,nyp)
```

In this solution the derivative operators Dx and Dy have been calculated at the volume centers (nxc and nyc) and then applied (dot) to a projection of the rhs from the mesh to the volume. The Boundary conditions are applied to the Dx and Dy operators as Dirichlet Boundary Conditions, which means that the velocity (in this case zero) at the boundary is kept constant. Alternative BC, e.g., *free slip*, as for the streamline solution, can be applied as well. See Problem 10.6. The Laplacian has also *no-slip* BC applied to it.

Using the same two anomalies as in Sect. 9.6 I obtain here the solution in Fig. 10.1. An immediate look shows that the streamline solution differs only near the walls, where the BC are in this case *no-slip* (we will see in a later section how to implement the other boundary conditions). However looking at the pressure field, also illustrated in Fig. 10.1, shows that something went wrong. A *checkerboard* pattern appears. This is a very general type of instability, that pops up at nearly every attempt to solve two or more equations on one grid. To understand why this happens is very important in order to build a correct multivariable solver and obtain a correct pressure solution.

From a purely numerical point of view, what happens with the pressure solution is that two incorrect solutions on the same grid can superpose one to each other in order and still solve the coupled system of equations by carefully compensating each other on *alternating* nodes. In other words if I would solve the equation on the even nodes and uneven nodes separately, I would obtain two solutions, one shifted from the other, but both solving my discretized equations. There is a large wealth of techniques for fixing this problem (see [98] for a review), but the most common one is to solve the momentum equation and the continuity equations on different grids.

In our case, the strategy is to solve the velocities on the nodes and the pressure at the center of each cell. Doing so requires to change our matrix, since the number of cells is $nxc \times nyc$, which is $nyp + nxp - 1$ less than the number of nodes $nxp \times nyp$. We have, therefore, to solve our system for $2 \times nxp \times nyp$ velocity unknowns and for $nxc \times nyc$ pressure unknowns.

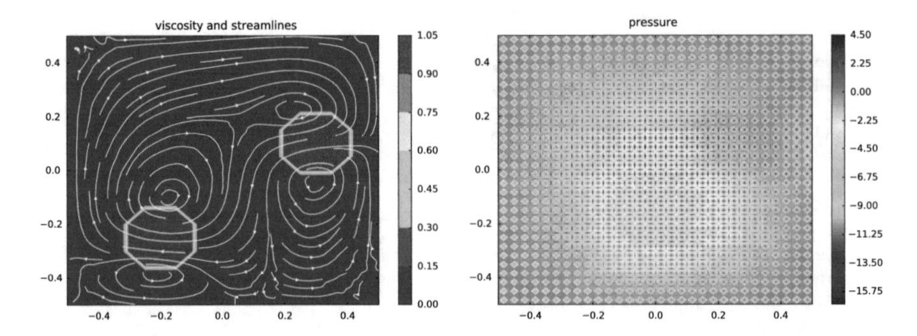

Fig. 10.1 *Left* velocity field determined with a full momentum+pressure solver for the same two particles in a box, using the full Stokes operator. Here, the solution is the same as the one in Fig. 9.5 except near the walls, due to the different Boundary conditions (free slip in the other case, no-slip in this case). *Right* pressure field for the same case. One observes that an instability arises. It arises because velocity and pressure are both solved on the same nodes

10.3 The Finite Volume Method

At the first sight, it might seam that the origin of the instability in calculating pressure illustrated in Fig. 10.1 lies in some minor error done for solving these equations. Instead in the side averaging used to calculate the gradient at the center of the cell. This strategy was in fact the best to minimize the error by calculating the x-derivative of v_x and y-derivative of v_y, however it suffers of information reduction. In fact while calculating the gradient of a generic field on the nodes of a lattice at the cell center, we are automatically reducing the number of points in which the information is defined from $nx \times ny$ to $(nx - 1) \times (ny - 1)$.

The appearance of the instability in the calculation of the pressure has emerged to computational fluid-dynamicists as early as the seventies years of the twenty centuries and many solutions have been put forward. It is possible to show that the oscillatory instability is a form of *over-fitting*, for which the most common remedy is regularization. Numerically regularization is implemented by using the *Penalty* method, i.e. a perturbation to the zero terms of the operator in (10.5). This technique is however quite complex, therefore I will show here a simpler one, based on discretizing the three equations in three different locations of the numerical domain.

This approach was one of earliest successful solutions to the pressure stabilization problem and was introduced by Patankar and Spalding ([30, 31]), who developed the SIMPLE algorithm. This consisted in calculating using the correct velocity solution to calculate a correction term to the velocity solution, and iterate until convergence was reached. This technique allows achieving the solution very rapidly and could be implemented with the operator approaches introduced in the Chap. 8. It is proposed as an advanced exercise (Problem 10.6). More detailed information on how to implement a generic finite volume discretization are in classic textbooks such as [47] and [48], or modern implementations applied to the geosciences such as [57].

The *Finite Volume Method (FVM)* is also a more physical way than the Finite Differences to understand the how we are solving our equations. Although ultimately the discrete implementation of the Finite Volume method is very similar (and in some cases identical) to the one emerging from the Finite Difference method, the Finite Volume approach is based on an integral in the volume of the equations.

FVM is based on an equilibrium law for an entire cell, not in every node. We have already seen in Chap. 8 that it is physically more sound and numerically more reliable to solve the continuity equation at the center of a cell. From Fig. 8.1 one observes that the difference between the velocities calculate at the center of each side are the best representation of the flux through the cell. This is another way to understand the importance of having a operator that average the values of v_x and v_y half cell in the x and y directions, respectively.

The implementation of the Finite Volume is simpler for a more compact equation, therefore I will show first how to discretize the pressure operator of the x-momentum equation, $\frac{\partial}{\partial x} p$. The indexes refer to the one of Fig. 10.2:

$$\frac{p_{i+\frac{3}{2},j+\frac{1}{2}} - p_{i+\frac{1}{2},j+\frac{1}{2}}}{\Delta x} \tag{10.6}$$

where the meaning of the half indexes, $\frac{1}{2}$ indexes are a simple way in which mid-points are described in the Finite Volume/ Finite Difference literature. It is important to observe that this term is calculated in the point $i+1, j+\frac{1}{2}$ and this is where we will have to calculate all the other terms. Let's look now at the second term of the x-momentum equation $\frac{\partial}{\partial y}\left(\eta\frac{\partial}{\partial x}\right)v_y$:

$$\frac{\eta^{i+1,j+1}\left(v_y^{i+\frac{3}{2},j+1}-v_y^{i+\frac{1}{2},j+1}\right)-\eta^{i+1,j}\left(v_y^{i+\frac{3}{2},j}-v_y^{i+\frac{1}{2},j}\right)}{\Delta x\Delta y} \tag{10.7}$$

where the key point is that the viscosity is calculated in the point where the intermediate derivate which is also centered in $i+1, j+\frac{1}{2}$. And finally the more complex term $\frac{\partial}{\partial x}\left(2\eta\frac{\partial}{\partial x}\right)+\frac{\partial}{\partial y}\left(\eta\frac{\partial}{\partial y}\right)v_x$:

$$2\frac{\eta^{i+\frac{3}{2},j+\frac{1}{2}}\left(v_x^{i+2,j+\frac{1}{2}}-v_x^{i+1,j+\frac{1}{2}}\right)-\eta^{i+\frac{1}{2},j+\frac{1}{2}}\left(v_x^{i+1,j+\frac{1}{2}}-v_x^{i,j+\frac{1}{2}}\right)}{\Delta x^2}+ \tag{10.8}$$

$$\frac{\eta^{i+1,j+1}\left(v_x^{i+1,j+\frac{3}{2}}-v_x^{i+1,j+\frac{1}{2}}\right)-\eta^{i+1,j}\left(v_x^{i+1,j+\frac{1}{2}}-v_x^{i+1,j-\frac{1}{2}}\right)}{\Delta y^2} \tag{10.9}$$

Where again the position where the viscosities are taken is naturally indicated in the Fig. 10.2. It is not necessary to write the formulation in the y-direction as it is perfectly symmetrical to the x-direction. The discretization of the continuity equation, that closes the system, has been already extensively analyzed in other chapters.

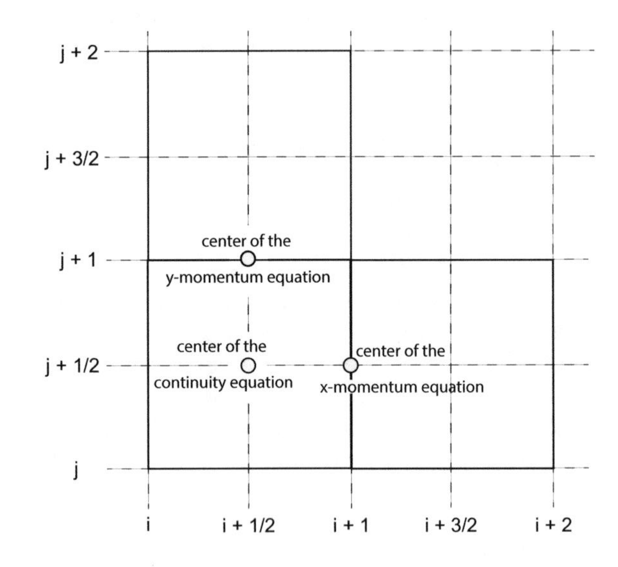

Fig. 10.2 Sketch of the of the indices and of the positions where the continuity, the x-momentum and the y-momentum equations are calculated. The mid indexes ($\frac{1}{2}$ and $\frac{3}{2}$) indicate the center between two edges of the volume. By calculating the finite volume at the center of each side one is also implicitamente integrating the momentum equations on each side. It is possible to show that this choice makes this algorithm second-order accurate.

10.4 Implementation of the Nonhomogenous Stokes Equation

Let's look at how a possible implementation of the algorithm just described can be implemented efficiently an in a completely vectorized fashion. One could build the matrix equivalent to the above formulation using the operators introduced in the past chapter, however due to the complications of calculating the numerous viscosities in different points of the domain, in this case it is possible to reconstruct the entire matrix in once, one submatrix at a time. Please, notice how kinemtic boundary conditions are set only to the bottom (iy=0) and left (ix=0) of the box. There are clearly many other options, primarily period boundary conditions that are often employed.

```
def fullStokesOperator(viscosity,dx,dy,nx,ny):
    nxy=nx*ny

    indexes = np.arange(nxy)
    XMinInd = (indexes<ny);
    XMaxInd = (indexes>=nxy-ny)
    YMinInd = (indexes % ny == 0)
    YMaxInd = (indexes % ny == ny-1)

    # Viscosity shift for momentum equation
    eta = viscosity.reshape(nxy) # stretch the array
    etaXmom = np.zeros(nxy); etaXmom[~XMinInd] = eta[~XMaxInd];
    etaYmom = np.zeros(nxy); etaYmom[~YMinInd] = eta[~YMaxInd];

    # Viscosity harmonic averages at cell center
    ind1 = (~XMinInd)*(~YMinInd); ind2 = (~XMinInd)*(~YMaxInd)
    ind3 = (~XMaxInd)*(~YMaxInd); ind4 = (~XMaxInd)*(~YMinInd)
    etaHarmonic = (eta[ind1]*eta[ind2]*eta[ind3]*eta[ind4])**0.25

    # cell or cell-side centered averages
    etaCell = np.ones(nxy); etaCell[ind1]=etaHarmonic;
    eta_XP =  np.ones(nxy); eta_XP[ind4]=etaHarmonic;
    eta_YP =  np.ones(nxy); eta_YP[ind2]=etaHarmonic;

    ##### x-momentum #####
    ind = (~XMinInd)  # exclude x==0

    # 2 d(eta dvx/dx)/dx + d(eta dvx/dy)/dy
    offsets=[ny,1,0,-1,-ny];
    diags=np.zeros((5,nxy),float)
    diags[0,ind] = +2.0 * etaXmom[ind]/dx**2
    diags[1,ind] = +1.0 * etaCell[ind]/dy**2 #-1
    diags[2,~ind] = 1.0;  # for x==0 vx velocities are imposed
    diags[2,ind] = -2.0 * (eta[ind]+etaXmom[ind])/dx**2 #2(d2vx/dx2)
    diags[2,ind] += -(etaCell[ind]+eta_YP[ind])/dy**2 #d2vx/dy2
    diags[3,ind] = +1.0 * eta_YP[ind]/dy**2  #+1
    diags[4,ind] = +2.0 * eta[ind]/dx**2
    Mx_vx = sparse.dia_matrix((diags,offsets), shape=(nxy,nxy)
  ↪ ).transpose().tocsr()

    # d(eta dvy/dx)/dy
    offsets=[-1+ny,-1,+ny,0];
    diags=np.zeros((4,nxy),float);
    diags[0,ind] = -eta_YP[ind]/dx/dy
```

```
    diags[1,ind] = +eta_YP[ind]/dx/dy
    diags[2,ind] = +etaCell[ind]/dx/dy
    diags[3,ind] = -etaCell[ind]/dx/dy
    Mx_vy = sparse.dia_matrix((diags,offsets), shape=(nxy,nxy)
↪   ).transpose().tocsr()

    # dp/dx
    offsets=[0,+ny];
    diags=np.zeros((2,nxy),float)
    diags[0,ind] = -np.ones(nxy)[ind]/dx
    diags[1,ind] = +np.ones(nxy)[ind]/dx
    Mx_p = sparse.dia_matrix((diags,offsets), shape=(nxy,nxy)
↪   ).transpose().tocsr()

    ##### y-momentum #####
    ind = (~YMinInd)   # exclude y==0

    # d(vx/dy)/dx
    offsets=[-ny+1,-ny,+1,0];
    diags=np.zeros((4,nxy),float)
    diags[0,ind] = -eta_XP[ind]/dx/dy
    diags[1,ind] = +eta_XP[ind]/dx/dy
    diags[2,ind] = +etaCell[ind]/dx/dy
    diags[3,ind] = -etaCell[ind]/dx/dy
    My_vx = sparse.dia_matrix((diags,offsets), shape=(nxy,nxy)
↪   ).transpose().tocsr()

    # (d2 vy/dx2) + 2 (d2 vy/dy2)
    offsets=[ny,1,0,-1,-ny];
    diags=np.zeros((5,nxy),float)
    diags[0,ind] = +1.0*etaCell[ind]/dx**2
    diags[1,ind] = +2.0*etaYmom[ind]/dy**2 #-1
    diags[2,~ind] = 1.0
    diags[2,ind] = -2.0 * (eta[ind]+etaYmom[ind])/dy**2 #2(d2vx/dx2)
    diags[2,ind] += -(etaCell[ind]+eta_XP[ind])/dx**2 #d2vx/dy2
    diags[3,ind] = +2.0*eta[ind]/dy**2   #+1
    diags[4,ind] = +1.0*eta_XP[ind]/dx**2
    My_vy = sparse.dia_matrix((diags,offsets), shape=(nxy,nxy)
↪   ).transpose().tocsr()

    # dp/dy
    offsets=[0,+1];
    diags=np.zeros((2,nxy),float)
    diags[0,ind] = -np.ones(nxy)[ind]/dy
    diags[1,ind] = +np.ones(nxy)[ind]/dy
    My_p = sparse.dia_matrix((diags,offsets), shape=(nxy,nxy)
↪   ).transpose().tocsr()

    ##### continuity #####
    ind = ~(XMinInd * YMinInd)

    # dvx/dx
    offsets=[0,-ny]
    diags=np.zeros((2,nxy),float)
    diags[0,ind] = -np.ones(nxy)[ind]/dx
    diags[1,ind] = +np.ones(nxy)[ind]/dx
    P_vx = sparse.dia_matrix((diags,offsets), shape=(nxy,nxy)
↪   ).transpose().tocsr()
```

```
    # dvy/dy
    offsets=[0,-1]
    diags=np.zeros((2,nxy),float)
    diags[0,ind] = -np.ones(nxy)[ind]/dy
    diags[1,ind] = +np.ones(nxy)[ind]/dy
    P_vy = sparse.dia_matrix((diags,offsets), shape=(nxy,nxy)
↪    ).transpose().tocsr()

    # identity for pressure at one corner
    P_p = sparse.coo_matrix((([1]),([0],[0])),shape=(nxy,nxy)).tocsr()

    return sparse.bmat([[Mx_vx, Mx_vy, Mx_p], [My_vx, My_vy, My_p],
↪    [P_vx,P_vy,P_p]]).tocsr()
```

The function `fullStokesOperator` is entirely vectorized, therefore every single operation can be optimized as any linear algebra operation. The function is longer than others that we analyzed before, because it combines the calculation of 9 sub-matrices in one routine. A quick look, however, shows how the same strategy is applied several times through its entire implementation. One could split it in smaller pieces (and it would be a useful exercise for the reader to do), but we can analyze it in detail and see that its overall structure not difficult to grasp:

A. XMinInd, XMaxInd, YMinInd, YMaxInd are arrays of indexes that comprises one of the four *edges* of the domain, i.e., minimum and maximum x and y. In this way one can immediately point to the remaining nodes with ∼*variable*.
B. Viscosity averages are either shifted (for the momentum equation) or a cell-centered harmonic average is calculated. This helps dealing with sharp viscosity variations. Once averaged, using indexes one has three types of averages: (i) cell centered; (ii) along x, (iii) along y.
C. Second derivatives are built using the traditional $a_{n-1} - 2a_n + a_{n+1}$ formulation, however the coefficient of the three values are weighted in function of viscosity, by combining the averages calculated in the domain.
D. Each of the nine sub-matrices has been built using the *dia_matrix()* function of *NumPy.sparse*. This allows fast and efficient matrix building, minimization of memory occupation and guarantees that the final matrix is the smallest possible.
E. Every submatrix is transposed when the *dia_matrix()* is created. This is due to the fact that the *offset* in building sparse matrices in NumPy shifts the diagonal only up and down, not laterally. Transposing the matrix overcomes this limitation in the Python library itself.

The most critical part of the routine is the careful averaging of the viscosity. The reader is invited to explore other averaging strategies. This is due to the presence of the derivatives of the viscosity in the full Stokes formulation and their apparent divergence when the viscosity gradients become extremely steep.

There are many alternative to this matrix construction. For example, it is possible to introduce ghost nodes (e.g., in [11]), and introduce a variety of different boundary conditions. However this goes beyond the scope of this text. Let us look instead now at how to combine this powerful routine with scaling instead the efficiency of this code on several setup and its scaling with problem size. We test to start with the flow relative to a sphere immersed in a fluid in a squared domain.

10.5 Long-Range Interaction

Geodynamics span through the scales, from the micro to the macro-scale. One of the main uses of numerical modeling is to allow extrapolating from small scale, heterogeneous media, an average behavior. Many particles that are highly packed in a fluid filled with a very viscous fluid experience a strong mutual force through the surrounding fluid. These particles can be strong or weak, but they will always interact. Since particle boundary have extremely sharp viscosity and density gradients, they are very at testing the validity of the finite volume implementation introduced in this chapter.

Let us start with a larger number of particles that the two that we used for the few tests above. For example with 20 or 30 particles randomly placed in a box, we should observe a collective dynamics. Will we observe that they all move as one? Or will they form clusters? Or move independently as rain drops? Let us write a simple routine that generates n hard spheres in a box, with a certain preimposed *fill* parameter that indicates what is the ratio of the volume occupied by the sphere. This is a simple exercise, a nice break from the heavy and difficult implementation of a solver. Still it is not trivial because we have to find a clever way to check that the spheres do not overlap and to re-randomly deploy the one of the two that overlaps. The following routine will do the job by building a distance matrix, using the *scipy.spatial* library.

```
def deploySpheres(sNumber,fill,minDist,xMin,yMin,xTot,yTot):
    sRadius = (fill*xTot*yTot/sNumber/np.pi)**0.5
    sCenters = np.arange(sNumber*2).reshape(sNumber,2)*1e-5

    farSpheres = False #True when non overlapping spheres
    while not farSpheres:
        k = distance.squareform(distance.pdist(sCenters))
        recCenters = np.where((k<2*sRadius+minDist) & (k>1e-6))[0]
        sCenters[recCenters]=np.random.rand(len(recCenters),2)
        sCenters[recCenters] *= (xTot-2*sRadius,yTot-2*sRadius)
        sCenters[recCenters] += (xMin+sRadius,yMin+sRadius)
        farSpheres = not len(recCenters)

    return(sCenters,sRadius)
```

With this routine we can create a simple model that simulates the evolution of, for example, 40 very stiff particles that occupy 10% of the volume (i.e. small particles distributed around). In order to impose stiffness within the inhomogeneous formulation above, we have to impose a viscosity two orders of magnitude greater than the surrounding fluid.

The evolution of this setup is shown in Fig. 10.3, on the left the density and the streamlines, and on the right the pressure (that does not anymore experience the checkerboard instability). Here, I show 30 steps where it is clear how the strong particles organize in convective cells. We therefore show that the flow of the spheres is not controlled by the buoyancy of each sphere, but by the strong interaction between

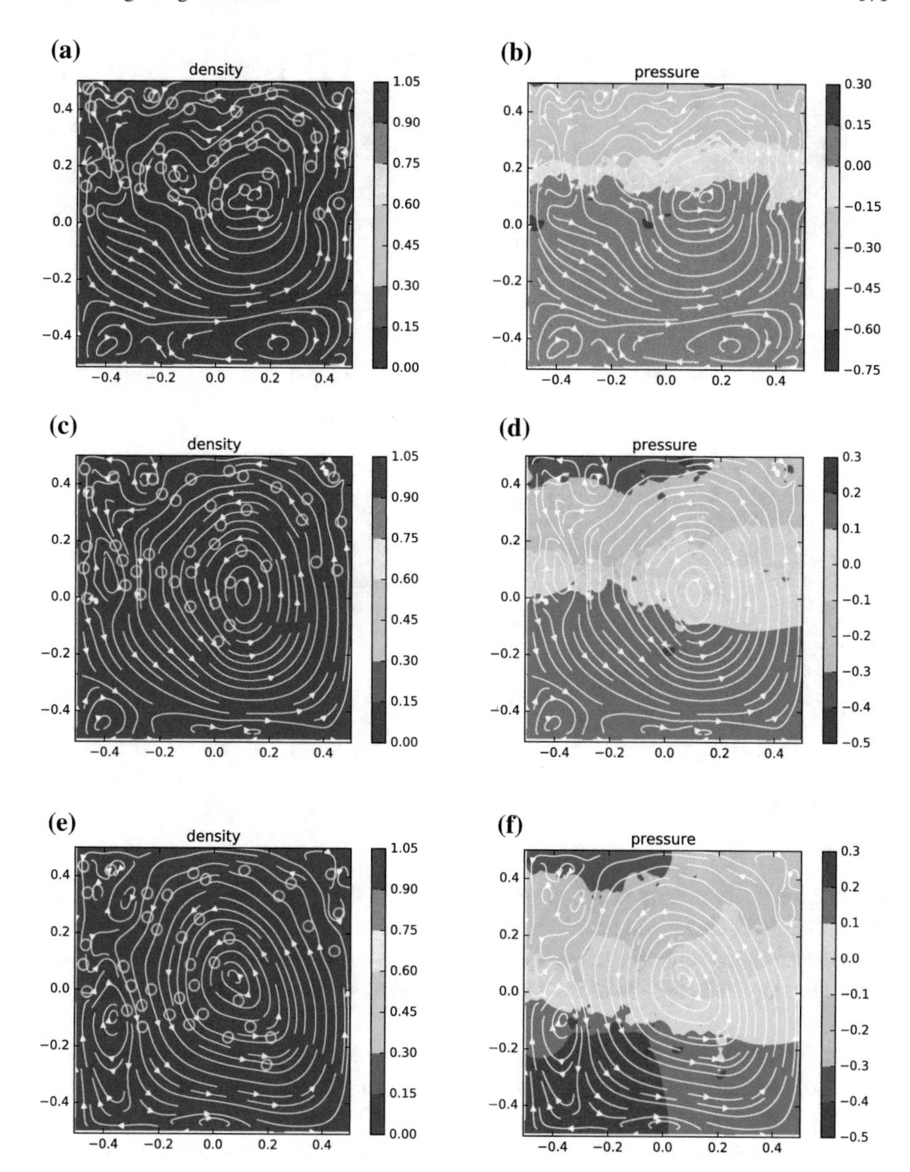

Fig. 10.3 Simulation of the convection induced by 40 spherical hard sphere (in 2D, so they are in fact cylinders) here modelled as inclusions Newtonian viscosity 100 times greater than the remaining homogeneous fluid. Left: density plot and streamlines. Right: pressure plot and streamlines. The spheres are placed randomly with a size calculated in order to fill 10% of the top portion of the domain. The snapshots represent the initial condition and the state each 50 steps. The resolution of the background lattice where the momentum equation is solved is 150×150. In each cell an average of 100 particles track the composition. The hard sphere tend to interact at large distance, as large as the entire domain, which creates a long scale cluster that transport almost all the hard spheres. As a consequence some sphere first rise and then fall with the others

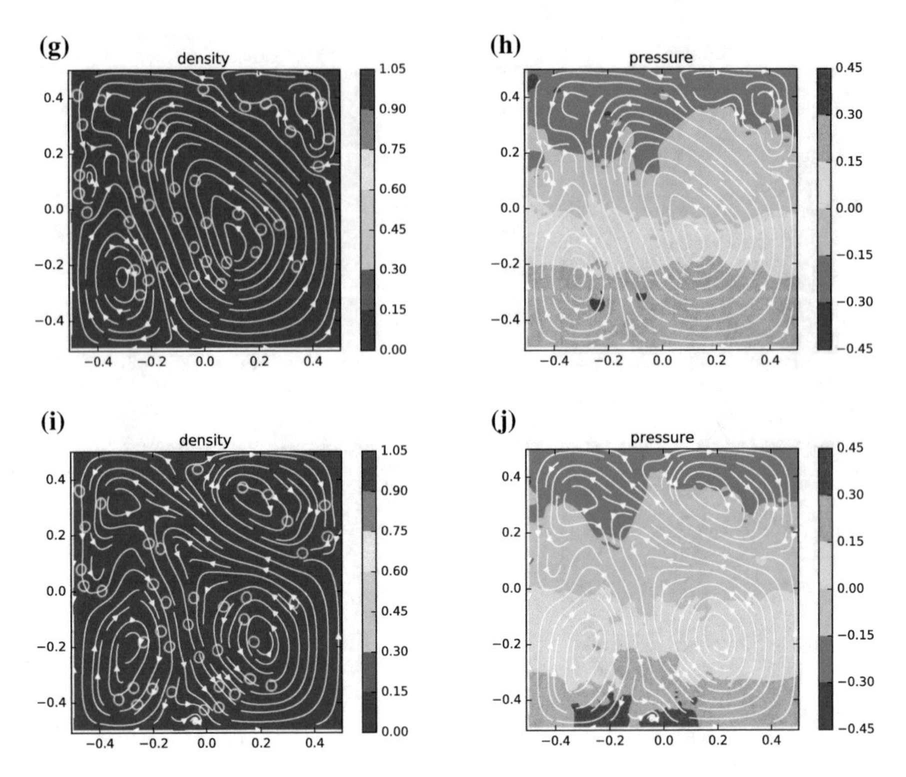

Fig. 10.3 (continued)

them. This phenomenon is called spheres in the fluid is controlled not by the runs of this model will show you how you can switch from a two-cell to one-cell behavior, with the particles self-organizing on or several convective pattern. Such stiff particles are an example of crystals in the magma as well as of platelets in the mantle.

On the opposite with this approach one can model a large number of very weak inclusions, analogue to bubbles in magma, or to partially melted regions in the mantle. A sample of this set of models is shown in Fig. 10.4, starting from the same setup of the model with rigid particles. While rigid particles are transported by the flow, here the weak regions deform and control the flow. For both setups, however, the regions seem to act as a collective, organized whole. This kind of organization is very interesting, common in geodynamics, as well as in many other fields of the physics of complex systems.

The greater the number of the particles, the better are resolved the boundaries of the spheres but also the more computationally demanding are the calculations of the advection and particles to mesh, and mesh to particles routines. In order to find the balance between the two it is essential to estimate the computational cost of each operation. On one or few processors this can be simply done by using the iPython magic keyword *%timeit*, as already employed in Problem 7.4. Ideally one wants to

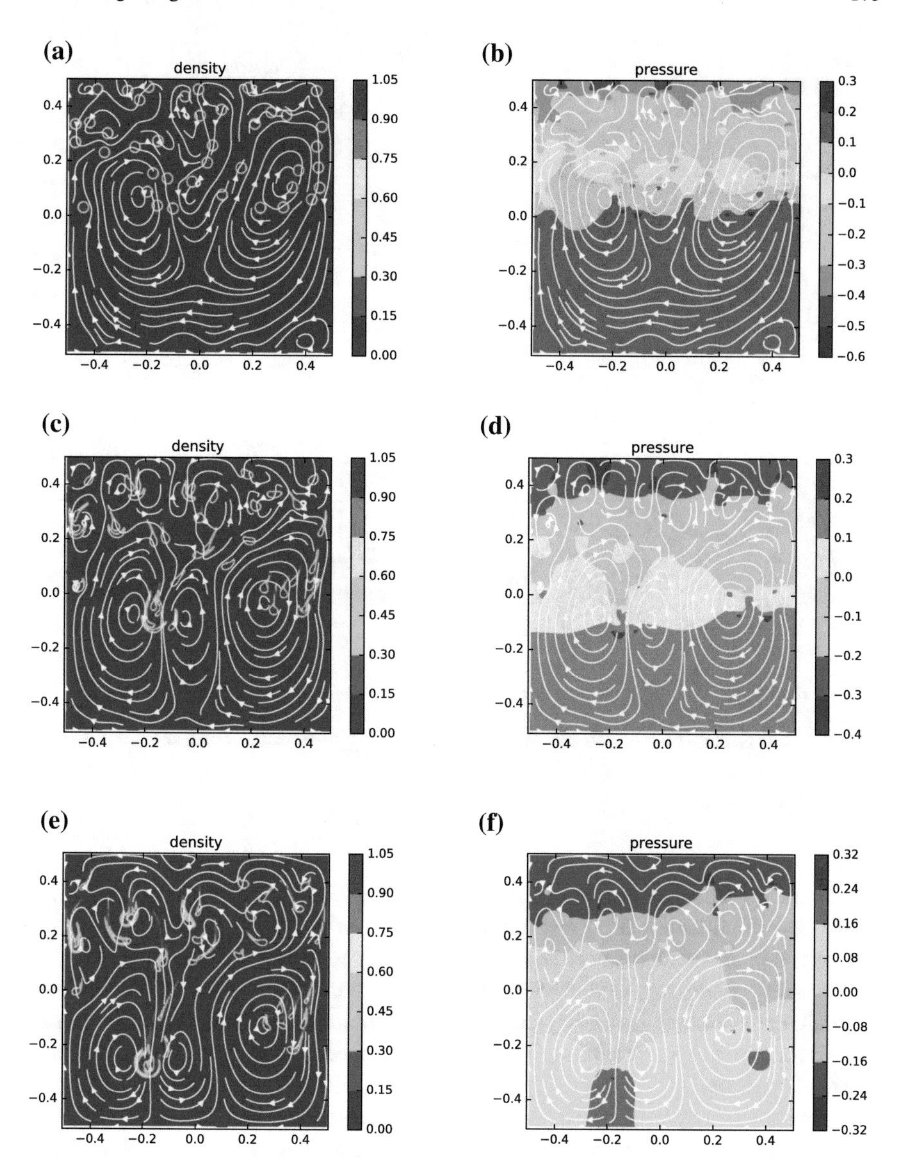

Fig. 10.4 Two-dimensional simulation of the convection induced by 40 initially spherical weak anomalies, here inclusions with a viscosity 100 times the background viscosity. The spheres are initially placed randomly in order to occupy 10% of the half portion of the domain. Five snapshots represent the initial condition and the state each 50 steps. The resolution of the background lattice is 150×150, with 100 cells on average in every cell. On the left it is shown the density, while on the right the pressure. Weak spheres, contrary to the hard ones of Fig. 10.3, tend to form small clusters, as large as few times the size of the initial anomaly, and precipitate independently in small groups where the sphere coalesce (surface tension is not present)

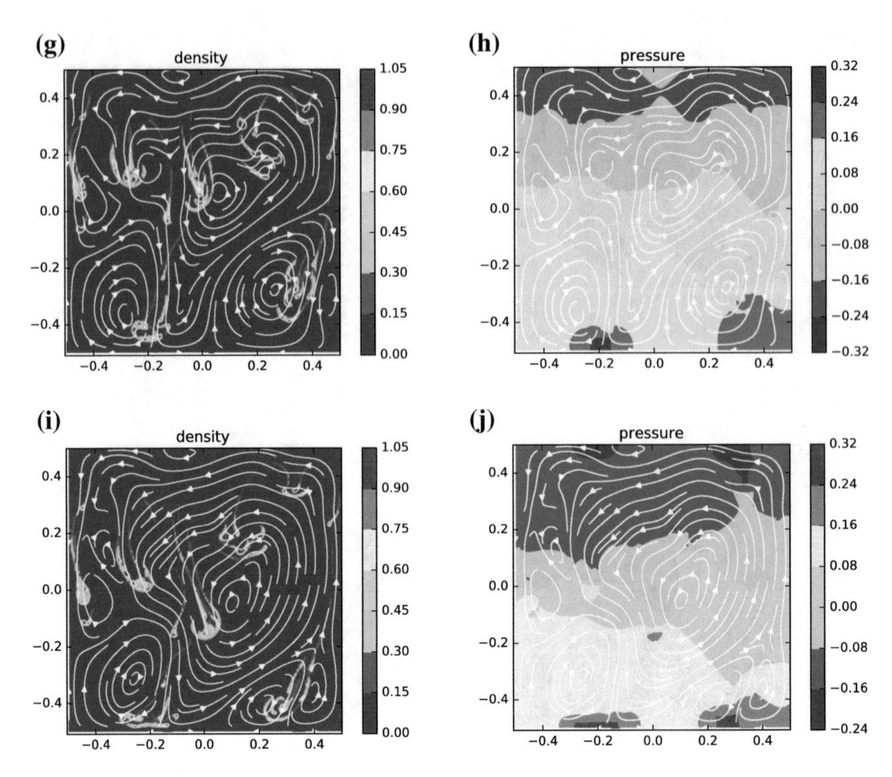

Fig. 10.4 (continued)

increase the number of particles until one reaches either the machine memory limits (not that difficult on a high resolution model) or when the particle operations all together demand 50% or more of the computing time for the entire simulation.

On many processors (hundreds up to tens of thousands) things can be ever more complicated. *%timeit* is not anymore a viable tool and it is better to monitor computing time on the fly by repeatedly calling the function *time()* of the module *time* at the root node and by subtraction calculate and store the time required for each major operation. Clearly, also each node can be monitored separately for more fine tracking of the time lag and other performance. It is important to take into account that on a large system Python codes need to be precompiled before running. This is because most MPI clusters use shared file systems such as NFS or Lustre, which implies that if a Python is running separately on every node and is calling the libraries from each of them at the same it, it will overload the file system potentially causing huge delays.

A key element of the models with sharp boundaries, like the advection of rigid particles, or of weak bubbles, is that they are based on a very accurate tracking algorithm. In fact the use of particles as tracers was essential to accurately describe the boundaries between different regions. Let us look now at a problem where the boundaries between domains with different density and viscosity are smoother and a less demanding advection scheme is sufficient.

10.6 Advection–Diffusion Equation

Due to the very long time required by its processes, most geodynamic phenomena do not display just advection of a material or field, but also its diffusion. There are several potential implementations of the advection and diffusion of a field. The easiest is probably to add an advection term to the standard Eulerian formulation of the diffusion equation.

Let us discretize both the velocity field and the first derivative of the field, that for simplicity and without loosing generality we will assume here to be Temperature. The diffusion term can be either expressed explicitly or implicitly, as extensively discussed in Chap. 9. It is generally recommended to use the implicit one, since it is intrinsically stable, however for every problem it is useful to calculate the stability criteria and if an explicit formulation is acceptable, it is recommended since it sensibly accelerates the calculations.

A possible one-dimensional Finite-Difference expression for implicit diffusion (with constant diffusivity) and implicit advection is the following:

$$\frac{T^{t+\Delta t}(x_i) - T^t(x_i)}{\Delta t} = k\frac{T^{t+\Delta t}(x_{i+1}) - 2T^{t+\Delta t}(x_i) + T^{t+\Delta t}(x_{i-1})}{\Delta x^2} - v_x(x_i)\frac{T^t(x_{i+1}) - T^t(x_{i-1})}{2\Delta x} \tag{10.10}$$

Here we have used a stable but diffusive approximation for the gradient of the temperature based on the field value on the side nodes respect to x_i to increase stability. This problem can be numerically solved by modifying the implementation presented in Sect. 9.4: initialize T, calculate IA, start the loop where first one calculates the gradient of T and finally update the temperature adding the diffusive field to the advected one:

```
LaplacianOp=sparseLaplacianOperator(nxp,nyp,dx,dy)
LaplacianOp=addBC(LaplacianOp,nxp,nyp)*r*dx*dx
IA = sparse.eye(nxp).tocsc()+LaplacianOp

for thisStep in np.arange(steps):
    dTdx[1:nx-1] = 0.5*(T[0:nx-2]-T[2:nx])/dx
    T = la.spsolve(IA,T) - vx*dTdx
```

The 2D version of the Advection Diffusion equation follows closely. The full Finite-Differences expression is in fact straightforward and emerges from the modification of the one in Sect. 9.5:

$$\frac{T^{t+\Delta t}(x_i, y_i) - T^t(x_i, y_i)}{\Delta t} =$$

$$k\frac{T^{t+\Delta t}(x_{i+1}, y_i) - 2T^{t+\Delta t}(x_i, y_i) + T^{t+\Delta t}(x_{i-1}, y_i)}{\Delta x^2} - v_x(x_i, y_j)\frac{T^t(x_{i+1}, y_j) - T^t(x_{i-1}, y_j)}{2\Delta x} +$$

$$k\frac{T^{t+\Delta t}(x_i, y_{i+1}) - 2T^{t+\Delta t}(x_i, y_i) + T^{t+\Delta t}(x_i, y_{i-1})}{\Delta y^2} - v_y(x_i, y_j)\frac{T^t(x_i, y_{j+1}) - T^t(x_i, y_{j-1})}{2\Delta y}$$

It is very instructive to try to implement this algorithm in 2D using the same velocity tests (rigid motion and Bell's flow) introduced in Chap. 7. In fact, surprisingly, these approximations just using an Eulerian method, result very ineffective, and many purely numerical errors appear in the form of artificial propagating waves. The most standard way to get rid of these problems is to use the Upwind scheme introduced in Sect. 7.5. There the calculation of the derivative is adapted to the direction of the flow.

Summary

- The Stokes equations, as the combination of the continuity and momentum equations, can be expressed as a unique operator.
- Stokes equations can be implemented as using a combination of the operator formulations introduced in Chap. 8. While this gives a reliable solution of in the velocity domain, it returns an unstable oscillating solution of the pressure. The implementation of the staggered grid solves the pressure instability.
- A finite volume solution, which automatically calculates the momentum and continuity solution at the cell center, is the most straightforward solution of the full Stokes equation. It also naturally solves the variable viscosity case by calculating the harmonic average viscosity of every cell from its corners.
- An operator generator for the combined non-linear Momentum and Continuity equations can be created by a unique function, following a standard Finite Volume formulation.
- Stokes solver combining Particles in Cell and Non-linear Stokes solver can address a very wide range of key problems such as the role of long-range interaction in strong and weak inclusion in a viscous matrix (which applies to process in geoscience like rheology of magma and sedimentation), and advection-diffusion equation that applies to all the fields of geophysical fluid-dynamics (e.g. thermochemical mantle convection, planetary core evolution).

Problems

10.1 Write a *linearFullStokesOperator* python function that creates the operator that solves the full Stokes equation in 2D in function of nxp, nyp, dx and dy, calculating velocity and pressure for the case of constant viscosity, using the routines described in Sect. 10.2. Using the above solution, that can be written in a clear manner with only 6-7 lines of code.

10.2 In order to have free slip boundary conditions instead of no-slip the routine *addBC* has to be modified specifically for the gradient operators that apply to the velocity field. This can be done by imposing a Dirichelet BC for the normal velocity (zero) and Neumann BC for the velocity along the boundary (zero derivative normal to the wall). Program it relative to the constant viscosity solution described in Sect. 10.2. Use the streamline solution as a benchmark. Observe whether and when the checkerboard instability appears.

10.3 Modify the matrix building function shown in Sect. 10.4 in order to change the Boundary Conditions into periodic on the horizontal axis. Run the example in Sect. 10.5 with a wide box and periodic boundary conditions and observe the emerging wavelengths for strong inclusions.

10.4 Advanced Problem. A natural alternative to the calculation of the full matrix operator for v_x, v_y and p is the use of a *strain rate operator* with dimensions $2nxy \times nxy$, projecting the velocity field v_x, v_y into the non diagonal xy component. A challenging exercise is to write a function that composes the solution operator from this *strain rate operator*. One can use nested *sparse.bmat* commands to combine matrices of different shape.

10.5 Build the implementation of the convection of a thermally heated box from below with constant viscosity, using the streamline approach. Verify that the critical Rayleigh number is between 10^3 and 10^4 and try to get closer to the known result $Ra = 1708$ [22]

10.6 Advanced Problem. Implement the SIMPLE algorithm ([30, 31]), based on the calculation of the velocity solution with the technique in using the operator approach introduced in Chap. 8. The approach consists in solving first for velocities, then correcting the pressure solution (to avoid oscillations) and to iterate between velocity and pressure until convergence.

Part IV
Advanced Techniques

Here, it is shown how to implement a variety of non-mesh-based numerical approaches for solving continuum mechanics equations that appear in geodynamics. In particular, in the first chapter an entire section is devoted to tree codes, where the reader is guided through examples into how to implement trees using hashing. Finally, Boundary Elements are introduced and applied to multiphase flow.

The following chapter shows some applications in geodynamics of the techniques that have been introduced in this book, ranging from Subduction dynamics to volcano dynamics (gas-magma flow), from mantle convection to the stress strain due to the interaction of multiple faults.

Finally the last chapter wraps up every and illustrates some of the forthcoming applications in geophysica and geodynamics for which Python will protagonist, among them Big Data Analysis and Artificial Intelligence/Machine Learning.

Chapter 11
Trees, Particles, and Boundaries

"Bad programmers worry about the code. Good programmers worry about data structures and their relationships."
— Linus Torvalds

Abstract This chapter address two types of advanced topics. One is how to build Trees. Two strategies are illustrated, (i) the Warren and Salmon implementation of the Barnes and Hut tree, based on the Hashing technique, and (ii) the standard KD tree, already implemented in the SciPy Libraries. The second topic of the chapter is how to build a Boundary Element solver for fluid-dynamics. The general theory of how to implement a multiphase flow is illustrated, and then an example of how to calculate the drag over a rigid body using quadratic triangular boundary elements is detailed with the specific solution for a flat ellipsoid. It is shown that a standard implementation of the BEM technique requires dense matrices, therefore scaling as N^2, where N is the number of the boundary Elements. To write a tree-based solver allows obtaining a fast solver that scales almost linearly with N.

In 1993, a revolution happened in the domain of Software Games. Most games that involved many players and dynamics were in two dimensions. When in 3D they were very slow. The game *Doom* changed the entire picture. Suddenly teenagers could play in first person in 3D moving in a dungeon at an amazing speed, from room to room. The breakthrough that brought to this revolution was the introduction of Binary Space Partitioning (BPS).

The idea behind it is the old *divide and conquer* algorithm. Instead of storing 3D information entirely, in Doom the space was represented by a binary tree that at every level zooms 2 times smaller cells. The leaves of the tree were convex polygons, without any further division. The way in which this algorithm allowed such a fast representation was to *traverse* the tree by starting at the root node and drawing the child nodes recursively, level by level. To decide which child node to draw first it was essential to know the position of the closest polygons to the camera, which had to be quickly accessed and drawn until the screen was filled, avoiding to draw objects that were not visible to the player.

G. Morra, *Pythonic Geodynamics*, Lecture Notes in Earth System Sciences,
DOI 10.1007/978-3-319-55682-6_11

This idea of using binary trees is present today in a large number of computer science algorithms and has become the standard way in which are modeled many body problems in Astrophysics (gravitational), Aerodynamics (turbulence), Electrostatics, and many more. Let us look now at how they appear in Geodynamics modeling. In the past section, we have seen how Stokes Flow (and so many other physics problems that are characterized by a *potential*) can be described by the summation, or integration, of many singular solutions. In this chapter, we dive into the computational challenge of integrating these *characteristic solutions*.

The far field $1/r = r^{-1}$ decay of the Green Functions of Diffusion and Stokes equations has general analogues in physics in the gravity and electromagnetic potentials, which decay as r^{-1} and as e^{ikr}/r respectively. Systems controlled by these potentials, and forces, display what is normally called in physics a *long range interaction*, which means that the decay is sufficiently slow that the total many weak far sources will give a comparable contribution to few strong near sources. This makes these problems harder to solve efficiently because the far field component has to be calculated carefully. When this very general phenomenon, this all-to-all interaction, appears we speak about the *many body* problem.

Problems of this specie arose first in computational physics when trying to calculate the evolution of N celestial bodies. For example, how 100,000 stars would move in the space to form a galaxy. 100,000 (10^5) bodies that interact each with each other have 10,000,000,000 (10^{10}) interactions. In general the number of forces that should be calculated is N^2, if N is the number of bodies. In this chapter, I will show some examples of how these calculations can be approximated with a small error reducing the number of forces to calculate to a number of the order of $NlogN$. We will use trees to organize the sources and separating the near and far-field contributions.

11.1 Tree Building

Many implementations can be found in the literature on how to build trees that organize a large set of data. I will show only some structures that are particularly suited for implementation using NumPy or for which an efficient parallel implementation is well documented.

11.1.1 The Barnes and Hut Tree

Barnes and Hut defined the three-dimensional tree structure by the following properties:

A. The space where the model is calculated is a large cube, split into an octree of cubical sub-cells so that every cell has eight descendants equal in size and shape.

B. There is a maximum number m of particles contained in each terminal cell (=leaf). If a node has less or equal to m particles it is not further divided. We will consider only the case in which $m = 1$, in this way each leaf will contain either one or zero particles.

The way in which Barnes and Hut describe the construction of their tree is recursive, i.e., they start with an empty tree and add one body after the other, each time adding a new branch or leaf. This procedure can be obtained by finding the most refined element of the current tree which contains the body, and then adding the body directly to that element, either by refining it further or by simply inserting the body. This method is, however, not so suitable to be easily vectorized. This limitation can be overcome by using a Cython implementation, or by using a more modern algorithm, from Warren and Salmon, that intensively exploits the NumPy library, or by using the existing *KDTree* features added to the Spatial library of the standard SciPy Module.

11.1.2 The Warren and Salmon Solution

The simplest way to build a tree, whose structure reflects the position of a set of particles, is by dividing the space into smaller and smaller region until not more than one particle remains in one cell. In many problems of physical interest, the particles represent quantities that are not homogeneously distributed. For example *stars* in computational astrophysics or hydrated rocks in geodynamics subduction modeling. For these reasons, the structure of the tree in real cases is highly heterogeneous. In 1993 Michael S. Warren and John K. Salmon, two students at the University of California in Santa Barbara and at the California Institute of Technology, respectively, developed a simple and very elegant formulation to build and manage trees that performs well also with very clustered and heterogeneous particles distributions.

Several innovative ideas were in their paper. The first was to use *hashing* for accelerating tree building and to simplify parallelization. For example, if we have a number of points masses in the space, we can store the center of mass by averaging the center of masses of the daughters cells. The way this is handled in programming is by storing the addresses of each cell by *pointing* to *children* cells, along with data referring to that cells, summarizing the global effects of the daughters. This is a very complex structure and requires much programming to be organized and parallelized on a large system. Hashing is a simple way to overcome these complications. It is based on the extracting the unique tree structure from the x,y, z coordinates themselves.

Hashing is a key scheme that provides a uniform addressing mechanism for homogeneously placed data. It works simply by translating the x, y, and z coordinates into a binary form, and then to create a long integer number by alternating the digits (0 or 1) for each of the coordinates. E.g. if x in binary is 0110, y is 1001, and z 1010, the hashing key number will be 011.100.101.010, i.e., the key is 1834 in decimals, with integers of 12 digits. The main advantages of calculating the keys associated to each element is that we can now order all the elements because particles whose sorted addresses are near to each other are also near in space.

Let us look at an example, which always helps to understand techniques. I create here a routine that takes three x, y, z coordinates (renormalized to 1), transforms them in integer that indicates in which cell in a $2^{level} \times 2^{level} \times 2^{level}$ virtual mesh are, and then transforms the three numbers in a hash-key binary, merges the three hash keys in a unique index that is returned. In this way, we have a unique integer that we can use to order a large number of particles distributed in space.

```
def createHash(x,y,z,level):
    cellSize = 0.5**(level+1)

    # so all binaries have the same length
    x += 1; y += 1; z += 1
    xb=bin(int(x/cellSize))
    yb=bin(int(y/cellSize))
    zb=bin(int(z/cellSize))

    # indices from three because xb, yb, zb start with 0b1
    hb = "".join(xb[i]+yb[i]+zb[i] for i in range(3,level+3))

    return int(hb, 2)
```

Where we used two little tricks. One was to increase x, y, z of one so that the binary starts from one. This can be simply called with one set of numbers, e.g., :

```
x = np.random.rand()
y = np.random.rand()
z = np.random.rand()
level = 10

hashKey = createHash(x,y,z,level)
```

Using hash keys as a unique index in an n-dimensional space offers several advantages at the same time:

A. The ordering allows immediately to know which set of particles are near the one that we are analyzing (just the ones that have a keys just before and after the hash-key of the particle)
B. The ordering allows a perfect parallelization even for highly heterogeneous particle distribution (such as the one in Fig. 11.1). Given N processors and M particles, it is possible to elaborate the first M/N particles in the first processor, the second M/N in the second processor, and so on. In this way automatically near particles will be assigned to the same processor.

Both characteristics are important in most calculations because the majority of interparticle force, or potential, decays with distance. For this reason one needs to calculate more accurately the force (or potential) of the closer neighbors and one can approximate (and sometime even neglect) the contribution of the very far particles.

Hashing was initially introduced by Irene Gargantini, a physicist from Milan, in Italy, who worked for many years at IBM in Switzerland before moving to Canada where she became the first women to held a chair in a computer science department. She invented the Hash-Keys method for creating oct-tree much faster than the method used at that time, based on pointers. Since hashed trees occupy less memory and are faster to build, they are the ideal tool to use in parallel programming. Her initial paper, in 1982, [128] was followed by hundreds of works that applied her approach to all the parallel algorithms existing at her time.

Let's now see a simple example that illustrates why Hash-Keys are so powerful. We can create a set of few thousands random points in space and by using the *color* feature of *matplotlib scatterplot* illustrate how this represents a natural order of points in space:

```
import numpy as np
import matplotlib.pyplot as plt
from mpl_toolkits.mplot3d import Axes3D

nPoints=5000
x = np.random.rand(nPoints)
y = np.random.rand(nPoints)
z = np.random.rand(nPoints)
level = 10
hashKey=np.zeros(nPoints,dtype=np.int64)
for i in np.arange(nPoints):
    hashKey[i] = createHash(x[i],y[i],z[i],level)

fig = plt.figure()
ax = fig.add_subplot(111, projection='3d')
ax.scatter(x, y, z, c=hashKey, alpha=0.5)
```

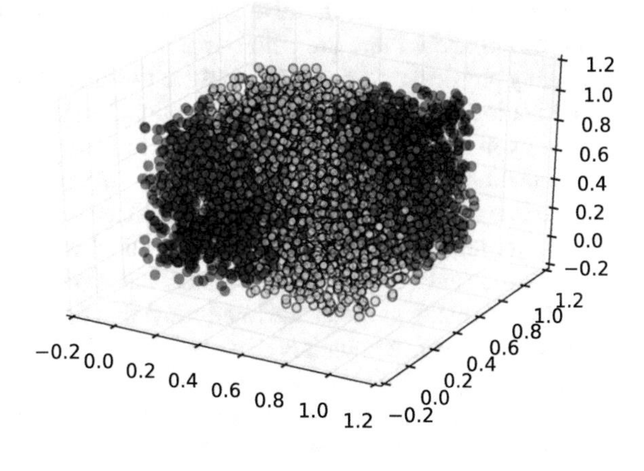

Fig. 11.1 Representation of few thousands points randomly placed in the space where the color represents the HashKey code associated with them. It is clear how the HashKey naturally represents a natural and elegant way to create a sequential organization of the points in the space that can be used for close interaction as well as dividing a domain in a cluster

The result is in Fig. 11.1. Clearly this implementation is very slow, as the Hash-Keys could be created in one with a unique vectorized operation. The implementation is left to the reader in one of the exercises at the end of the chapter.

11.2 SciPy k-d Tree

Binary trees have also been implemented into the Scientific Python Library (SciPy), within the *spatial* module and can be loaded with the instruction `from scipy import spatial`. This implementation allows space partitioning of data structure. The functions of this module are highly optimized and based on the standard k-d Tree Model. k stays for the number of dimensions in which we physical space exists. For example 3-d trees in three dimensions.

The idea behind *k-d Trees* is to split the domain in two at different dimensions for every level. In 3D, for example, at the first level the domain will be a split by an axis in the x direction, at the second level in the y direction, at the third level in the z direction, and then back to x. The axis and splitting point are chosen to be the *median* particle. Although the algorithm has been written for any dimension k, it is really efficient only at low dimensions (less than 10). To search for high-dimension nearest-neighbors is still an open scientific problem.

It is also possible to search for all-neighbors queries, although k-d trees are not very efficient this sort of calculation.

11.3 Boundary-Based Simulations

Most of this book has been devoted on how to accelerated codes that calculate the solution of the momentum or energy equation in the entire modeled space. This is the simplest and most intuitive approach to solving continuum mechanics equations, nevertheless there are better approaches for a large number of problems. In fact many physical equations that apply to the three-dimensional space are based the minimization or maximization of a functional. For example, every system characterized by many particles like a gas (made of many molecules) will tend toward the maximization of the entropy. Even systems far from equilibrium still have to obey to such laws. For example, the electrical properties of a metal is determined by the motion of many interacting electrons and will always tend toward the minimization of the associated quantum energy functional. Sometimes this results in surprising laws. For example, in optics the Snell's law that determines the incidence angle of a ray of light through a boundary between two system with different refractive indices ($n_1 \sin\theta_1 = n_2 \sin\theta_2$) is in turn the solution of the minimization problem: "what is the trajectory that minimizes the time necessary to go from the source to the target?" as brilliantly found by Fermat in 1662 [102].

Similarly for Stokes flow we have a principle that affirms that the flow solution of the momentum equation will be the one that minimizes energy dissipation [32]. By using this principle, it is therefore possible to write equations that are valid only at the boundary between two regions, each with an homogeneous density and viscosity characteristics. The first person who found the *Boundary Integral* expression for the multiphase viscous flow was Olga Ladyzhenskaya, a great soviet mathematician who first developed a great passion for mathematics from her father, a mathematics teacher, and then was blocked by the soviets who arrested and killed her father (while she was still a teenager) and prohibited her from studying math at the Leningrad University. It was only after the death of Stalin that she was allowed to present her PhD thesis and continue her career as a mathematician. She has left us with the first rigorous proof of the convergence of a finite difference method for the Navier-Stokes equations and for the first clear expression of the Boundary Element Method applied to Viscous Flow. Her major findings can be found in her famous book, [51]: *The Mathematical Theory of Viscous Incompressible Flow* from1969.

The theory behind boundary element is mathematically very advanced for this book, I will therefore embed only a synthesis in this gray box. Taking a domain where Stokes is the dominating velocity and the internal viscosity is η, the velocity at every point inside the domain D can be written as the sum of surface integrals of the velocity for a specific *kernel*. Kernels used in the Boundary Element Method are also called Green Functions. G_{ij} and T_{ijk} are the steady Green's functions (*Stokeslet* and *Stresslet*) for velocity and stress respectively that we introduced in Chap. 6.

The fundamental expression that relates the traction applied to the boundary is $f_i(\mathbf{x}) = \sigma_{ik}(\mathbf{x})n_k(\mathbf{x})$ with the velocity of every point at the boundary $u_i(\mathbf{x})$ is:

$$-\frac{1}{8\pi\eta}\int_{\partial D}\sigma_{ik}(\mathbf{x})n_k G_{ij}(\mathbf{x},\mathbf{x}_o)dS(\mathbf{x}) + \frac{1}{8\pi}\int_{\partial D}u_i(\mathbf{x})n_k T_{ijk}(\mathbf{x},\mathbf{x}_o)dS(\mathbf{x}) = \begin{cases} u_i(\mathbf{x}_o) & \text{if } \mathbf{x}_o \in D, \\ 0 & \text{otherwise} \end{cases} \quad (11.1)$$

where the integral of the traction is called *Single Layer* and the integral of the velocity *Double Layer*.

The expression in Eq. 11.1 was used by Ladyzhenskaya ([51], p. 75), to find the expression that relates velocity and stress on the boundary between two fluids with different viscosity. Formally, given a domain D and its boundary ∂D, with viscosity η inside ∂D and $\lambda\eta$ outside, we obtain inside ∂D:

$$u_j(\mathbf{x}_o) = -\frac{1}{8\pi\eta}\int_{\partial D} \sigma_{ik}(\mathbf{x})n_k^{in}(\mathbf{x})G_{ij}(\mathbf{x},\mathbf{x}_o)\, dS(\mathbf{x}) + \frac{1}{8\pi}\int_{\partial D} u_i(\mathbf{x})n_k T_{ijk}(\mathbf{x},\mathbf{x}_o)dS(\mathbf{x}) \tag{11.2}$$

and outside ∂D:

$$u_j(\mathbf{x}_o) = \frac{1}{8\pi\lambda\eta}\int_{\partial D} \sigma_{ik}(\mathbf{x})n_k^{out}(\mathbf{x})G_{ij}(\mathbf{x},\mathbf{x}_o)\, dS(\mathbf{x}) + -\frac{1}{8\pi}\int_{\partial D} u_i(\mathbf{x})n_k T_{ijk}(\mathbf{x},\mathbf{x}_o)dS(\mathbf{x}) \tag{11.3}$$

When the point x lies on ∂D a jump condition establishes that the two above equations become:

$$\frac{1}{2}u_j(\mathbf{x}_o) = -\frac{1}{8\pi\eta}\int_{\partial D} \sigma_{ik}(\mathbf{x})n_k^{in}(\mathbf{x})G_{ij}(\mathbf{x},\mathbf{x}_o)\, dS(\mathbf{x}) + \frac{1}{8\pi}\int_{\partial D} u_i(\mathbf{x})n_k T_{ijk}(\mathbf{x},\mathbf{x}_o)dS(\mathbf{x}) \tag{11.4}$$

$$\frac{1}{2}u_j(\mathbf{x}_o) = \frac{1}{8\pi\lambda\eta}\int_{\partial D} \sigma_{ik}(\mathbf{x})n_k^{out}(\mathbf{x})G_{ij}(\mathbf{x},\mathbf{x}_o)\, dS(\mathbf{x}) + -\frac{1}{8\pi}\int_{\partial D} u_i(\mathbf{x})n_k T_{ijk}(\mathbf{x},\mathbf{x}_o)dS(\mathbf{x}) \tag{11.5}$$

hence summing the two equations:

$$\frac{1+\lambda}{2}u_j(\mathbf{x}_o) = \frac{1}{8\pi\eta}\int_{\partial D} \Delta f_i(\mathbf{x})G_{ij}(\mathbf{x},\mathbf{x}_o)\, dS(\mathbf{x}) + -\frac{1-\lambda}{8\pi}\int_{\partial D} u_i(\mathbf{x})n_k^{out}(\mathbf{x})T_{ijk}(\mathbf{x},\mathbf{x}_o)dS(\mathbf{x}) \tag{11.6}$$

where $\Delta f_i(\mathbf{x})$ represents the jump in the traction between inside and outside the boundary $\Delta f_i(\mathbf{x}) = \sigma_{ik}^{out}(\mathbf{x})n_k^{out}(\mathbf{x}) + \sigma_{ik}^{in}(\mathbf{x})n_k^{in}(\mathbf{x}) = [\sigma^{out}(\mathbf{x}) - \sigma_{ik}^{in}(\mathbf{x})]n_k^{out}(\mathbf{x})$. An extensive literature about extrapolating the differential traction at boundaries for fluid-dynamic systems exist. When modelling systems with different buoyancy between inside and outside the boundary, one can simply write: $\Delta f\ (\mathbf{x}) = \Delta\rho\, \mathbf{g}\cdot \mathbf{x} n_i^{out}(\mathbf{x})$. If surface tension is present one has to add $\Delta f_i(\mathbf{x}) = \gamma n_i^{out}(\mathbf{x})$.

An intriguing characteristic of (11.6) is that the double layer appears only when the viscosity inside and outside the boundary ∂D is different. One can show how one obtains in this case the traditional Stokes equation for constant viscosity.

One can in fact write many more expressions for a multilayer, many systems with different viscosities, or extend the Boundary Element expression to elasticity (by a modification of the associated Green Function). But this goes beyond the scope of this book and interest contemporary research such as in [5, 93].

11.3.1 Drag over a Rigid Particle

In order to understand how the integral equations of Sect. 11.3 can be calculated numerically, let's start from the simplest case, the one of the drag over a rigid object. Since in this case the object does not deform, we need to calculate only the *Single Layer* integral to know the resistance.The discretized expression associated to the integral expression in (11.6) is called *The Boundary Element Method* and consists in dividing the surface of interest in elements, in our case triangles, calculate the integrals on each triangle separately, and then sum all of them. In the case of a rigid particle the viscosity inside the particle is virtually infinite, and therefore the *Double Layer* integral disappears:

$$u_j(\mathbf{x}_o) = \frac{1}{8\pi\eta}\int_{\partial D} \sigma_{ik}(\mathbf{x})n_k(\mathbf{x})G_{ij}(\mathbf{x},\mathbf{x}_o)dS(\mathbf{x}) \tag{11.7}$$

Where the positive sign in front of the integral indicates that the flow calculated is outside the domain. The drag over a rigid particles can be described with a 3×3 matrix, usually called *Grand Resistance Matrix* [32]. This matrix is symmetric and has the properties of a tensor, i.e. it depends on the particles characteristics and position, and not by the reference system. The assumption of rigidity is essential, as

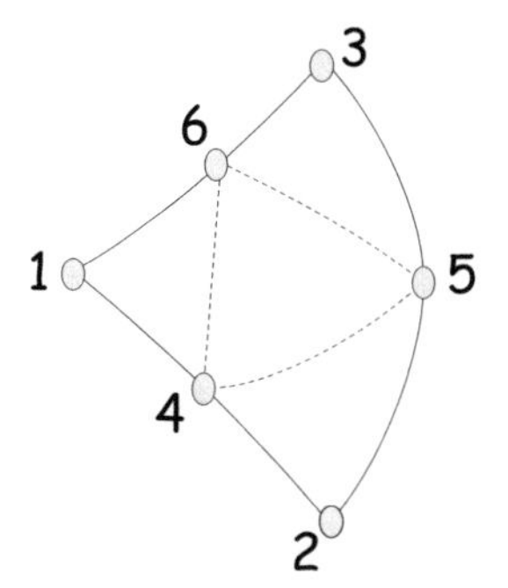

Fig. 11.2 Representation of a Quadratic Triangular Boundary Element that is used in the calculation of the integral over the surface of a particle. The total integral, in a quadratic approximation, is expressed by the sum of the linear integral on each of the four triangles of the figure. The singular integral of the *influence* of a triangle on itself is calculated by a loop summation over the six segments composing the boundaries of the triangle.

there is no single tensor able to describe the interaction between a deforming body and the surrounding fluid in Stokes flow.

As for the lattice based techniques, there are several *orders* of precision to calculate the integrals. The most general choices are (i) first order, (ii) second order, (iii) analytical integration. Also spectral techniques are often employed, particularly in seismology. I will show here how to calculate this integral using second order boundary element. This gives a much greater precision than first order for objects displayed a pronounced curvature, like particles immersed in a fluid. This is done by describing every surface triangle with 6 nodes, three at the triangle corners and three at the mid-face points, as described in Fig. 11.2. In a quadratic integration the integral over a full triangle is split in the four linear integrals shown in the figure.

Starting from a surface triangulation (e.g. the Fibonacci Covering introduced in Chap. 2), we need to use two routines to calculate the geometrical properties of the surface. The first related to the Construction of the *Second Order* Mesh, i.e. the placement of the six nodes for each triangle, and the second related to the calculation of Gauss positions necessary for the numerical integration over each triangle. The third routine relates to the calculation of the integrals themselves. Let's start from the geometrical features.

11.4 Quadratic Triangular Elements Mesh

A first routine is necessary to take a surface mesh and turn it into a quadratic mesh. The strategy is (i) calculate the mid points as an average of the triangular corners, (ii) project the mid points on the surface (a sphere in our case) in order to have a curved triangular elements, (iii) create a new array with all the nodes coordinates, and a new connectivity array in which six nodes are associated to each boundary element. Here follows a completely vectorized version of an implementation that starts from the Fibonacci Covering that we already seen in Chap. 2.

```
def createSecondOrderSphericalMesh(nPoints):
    (points,sphereMesh) = fibonacci_covering(nPoints)
    elNum = sphereMesh.simplices.shape[0] # number of elements
    sphereConn = sphereMesh.simplices #sphere connectivity

    #find mid points
    midPoints = np.zeros((elNum,3,3))
    midPoints[:,0]=0.5*(points[sphereConn[:,0]] +points[sphereConn[:,1]])
    midPoints[:,1]=0.5*(points[sphereConn[:,1]] +points[sphereConn[:,2]])
    midPoints[:,2]=0.5*(points[sphereConn[:,2]] +points[sphereConn[:,0]])

    #set the mid Points on the sphere (for another shape use normals)
    midPoints/=np.linalg.norm (midPoints,axis=2).reshape(elNum,3,1)

    # combine old and new points
    coordinates = np.append(points,midPoints.reshape(elNum*3,3),axis=0)
    pointsNumber = coordinates.shape[0]
```

```
# create the new connectivity array
conn = np.zeros((elNum,6), dtype=np.int) #connectivity
conn[:,0:3] = sphereConn
shift = points.shape[0]
conn[:,3] = np.arange(elNum)*3 + shift
conn[:,4] = np.arange(elNum)*3 + 1 + shift
conn[:,5] = np.arange(elNum)*3 + 2 + shift
return(coordinates,conn,elNum,pointsNumber)
```

Once the full mesh has been created, one has to identify the Gauss Positions necessary for the integration over each triangle, and also to calculate the normal vectors for every triangle, taking care of the mean element curvature. This can be done using standard procedures from geometry of curved surfaces.

The surface of each triangular element can be mapped into a more familiar right isosceles triangle. In this way one needs only two parameters ξ and η to calculate the integral.

$$\mathbf{x} = \sum_{i=1}^{6} \mathbf{x}_i \phi_i(\xi, \eta) \tag{11.8}$$

The detailed mapping procedure is described in textbooks such as [33], section 5.3. The idea is that derivatives over the original triangle can be mapped into derivatives over the right triangle. The six interpolation functions for our case are:

$$\phi_2 = \frac{1}{1-\alpha}\xi\left(\xi - \alpha + \frac{\alpha-\gamma}{1-\gamma}\eta\right)$$

$$\phi_3 = \frac{1}{1-\beta}\eta\left(\eta - \beta + \frac{\beta+\gamma-1}{\gamma}\xi\right)$$

$$\phi_4 = \frac{1}{\alpha(1-\alpha)}\xi\,(1-\xi-\eta)$$

$$\phi_5 = \frac{1}{\gamma(1-\gamma)}\xi\eta$$

$$\phi_6 = \frac{1}{\beta(1-\beta)}\eta\,(1-\xi-\eta)$$

$$\phi_1 = 1 - \phi_2 - \phi_3 - \phi_4 - \phi_5 - \phi_6$$

where α, β, γ are:

$$\alpha = \frac{1}{1+\frac{|x_4-x_2|}{|x_4-x_1|}} \qquad \beta = \frac{1}{1+\frac{|x_6-x_3|}{|x_6-x_1|}} \qquad \gamma = \frac{1}{1+\frac{|x_5-x_2|}{|x_5-x_3|}}$$

The calculation of these coefficients is long and tedious, but they can all be easily vectorized. Here follows an implementation:

```
def createGaussPositions
(allPoints,conn,elNum,xiCoord,etaCoord,gaussPoints):

gaussPositions = np.zeros((elNum,gaussPoints,3),float)
DDxi = np.zeros((elNum,gaussPoints,3),float)
DDeta = np.zeros((elNum,gaussPoints,3),float) for gaussPoint in
np.arange(gaussPoints):
    xi,eta = xiCoord[gaussPoint],etaCoord[gaussPoint]
    # six segments at the boundary of the quadratic triangular element
    d42 = np.linalg.norm(allPoints[conn[:,3]] -allPoints[conn[:,1]],axis=1)
    d41 = np.linalg.norm(allPoints[conn[:,3]] -allPoints[conn[:,0]],axis=1)
    d52 = np.linalg.norm(allPoints[conn[:,4]] -allPoints[conn[:,1]],axis=1)
    d53 = np.linalg.norm(allPoints[conn[:,4]] -allPoints[conn[:,2]],axis=1)
    d63 = np.linalg.norm(allPoints[conn[:,5]] -allPoints[conn[:,2]],axis=1)
    d61 = np.linalg.norm(allPoints[conn[:,5]] -allPoints[conn[:,0]],axis=1)

    alpha = 1.0/(1.0+d42/d41)
    beta = 1.0/(1.0+d63/d61)
    gamma = 1.0/(1.0+d52/d53)

    # calculate the basis functions coefficients for projecting xi and eta
    ph2 = xi *(xi-alpha+eta*(alpha-gamma)/(1-gamma))/(1-alpha)
    ph3 = eta*(eta-beta+xi*(beta+gamma-1)/gamma)/(1-beta)
    ph4 = xi *(1-xi-eta)/alpha/(1-alpha)
    ph5 = xi*eta/gamma/(1-gamma)
    ph6 = eta*(1-xi-eta)/beta/(1-beta)
    ph1 = 1-ph2-ph3-ph4-ph5-ph6

    # project from a surface element to a right triangle
    ph = np.array([ph1,ph2,ph3,ph4,ph5,ph6])
    coords = np.array(allPoints[conn[:,0:6]])
    for el in np.arange(elNum):
        gaussPositions[el,gaussPoint,0] = np.dot(coords[el,:,0],ph[:,el])
        gaussPositions[el,gaussPoint,1] = np.dot(coords[el,:,1],ph[:,el])
        gaussPositions[el,gaussPoint,2] = np.dot(coords[el,:,2],ph[:,el])

    # calculate the xi derivatives of basis functions
    dph2 = (2*xi-alpha+eta*(alpha-gamma)/(1-gamma))/(1-alpha)
    dph3 = eta*(beta+gamma-1)/gamma/(1-beta)
    dph4 = (1-2*xi-eta)/alpha/(1-alpha)
    dph5 = eta/gamma/(1-gamma)
    dph6 = -eta/beta/(1-beta)
    dph1 = -dph2-dph3-dph4-dph5-dph6

    # compute dx/dxi from xi derivatives
    dph = np.array([dph1,dph2,dph3,dph4,dph5,dph6])
    for el in np.arange(elNum):
        DDxi[el,gaussPoint,0] = np.dot(coords[el,:,0],dph[:,el])
        DDxi[el,gaussPoint,1] = np.dot(coords[el,:,1],dph[:,el])
        DDxi[el,gaussPoint,2] = np.dot(coords[el,:,2],dph[:,el])

    # evaluate eta derivatives of basis functions
        pph2 = xi*(alpha-gamma)/(1-alpha)/(1-gamma)
    pph3 = (2 * eta - beta + xi *(beta + gamma -1)/gamma)/(1-beta)
    pph4 = -xi/alpha/(1-alpha)
    pph5 = xi/gamma/(1-gamma)
    pph6 = (1-xi-2*eta)/beta/(1-beta)
    pph1 = -pph2-pph3-pph4-pph5-pph6
```

```
    # compute dx/deta from eta derivatives of phi
    pph = np.array([pph1,pph2,pph3,pph4,pph5,pph6])
    for el in np.arange(elNum):
        DDeta[el,gaussPoint,0] = np.dot(coords[el,:,0],pph[:,el])
        DDeta[el,gaussPoint,1] = np.dot(coords[el,:,1],pph[:,el])
        DDeta[el,gaussPoint,2] = np.dot(coords[el,:,2],pph[:,el])

    normalVector= np.cross(DDxi,DDeta)
    gaussMetrics = np.linalg.norm(normalVector,axis=2)

    return(gaussPositions,gaussMetrics)
```

This implementation is general, for the ξ and η gauss points passed to the routine. This implies that this implementation is vectorized over all the elements and run on any number of integration points. We will use it only for a specifically small number of integration points (4 for the non-singular integrals and 6 for the singular integrals) however it can be employed for any number of integration points over a right triangle.

11.4.1 Calculation of the influence matrix

The calculation of the influence matrix is at the core of this implementation of the Boundary Element Method. This is done by calculating the influence that each triangle has over every triangle, by integrating the Green function of the Stokes Equation between the two.

There are two possible integrals, one singular and one non-singular. Non-singular integrals appear when calculating the effect of the Green Function Integral (the Kernel) on a different integral, while singular integrations appear when one elements interact with itself. Since there are N_{el} boundary elements, there will be $N_{el}(N_{el}-1)$ non-singular integrals and N_{el} singular ones. Despite their small number, singular integrals are quantitively much greater than non-singular ones, therefore their precise calculation is very important. The singularity simply emerges from the Green Functions as their leading term is of the form $1/r$. Because the distance between two points on the same triangles will go to zero, the integral will have to take into account cancelling infinities. Following standard integration techniques as in the literature, singular integrals are performed using local polar coordinates (e.g. [33]) or analytically ([37]).

In the following implementation the strategy is to first calculate all the triangle-triangle integrations assuming that there are not singular ones. In this way a $N_{el} \times N_{el}$ matrix is filled with all the integrals. Then calculate the self-interaction one using polar coordinate strategy. The singular integrals require a particularly laborious set of operations, however they have to be done only once for each element, therefore that calculations scale linearly with the problem size. The calculation of the non-singular interaction integrals, instead, are a worrying N_{el}^2, which is the weak point of this specific implementation of the BE Method.

```
def calculateInfluenceMatrix (elNum,gaussPoints,gaussWeights,gaussPositions,
        legPoints,legWeights,legCoords,
        collocationPos,gaussMetrics,
        allPoints,conn,viscosity):

    influenceMatrix=np.array(np.zeros((elNum*3,elNum*3),float))

    # influence between boundary Elements using gauss integration
    for el in np.arange(elNum):
        for gaussPoint in np.arange(gaussPoints):
        d = gaussPositions[:,gaussPoint] - collocationPostion[el1]
        dd=np.einsum('ij,ik->ijk',d,d)
        r=np.linalg.norm(d,axis=1)[:,np.newaxis,np.newaxis]
        i = np.outer(np.ones(elNum), np.identity(3)).reshape(elNum,3,3)
        greenFunction = i/r + dd/r**3
        prefactor=0.5*gaussMetrics[:,gaussPoint]*gaussWeights [gaussPoint]
        influenceMatrix[el1::elNum,::] += np.einsum('i,ikl-
    >lki',prefactor,greenFunction).reshape(3,3*elNum)

        # influence between each triangle and itself using Lagrange integration
        for el in np.arange(elNum): influenceMatrix[el::elNum,el::elNum]=0.
        phi = np.pi/4.*(1.+legCoords)
        rMaxH = .5 / ( np.cos(phi) + np.sin(phi) )
        prefactorPhi = legWeights * rMaxH
        radius = np.outer(rMaxH,1+legCoords)
        xi = np.einsum('ij,i->ij',radius,np.cos(phi))
        eta = np.einsum('ij,i->ij',radius,np.sin(phi))
        zed = 1.0-xi-eta

    for index1,index2 in np.array([[0,3],[3,1],[1,4],[4,2],[2,5],[5,0]]):
        p0=collocationPos
        p1=allPoints[conn[:,index1]]
        p2=allPoints[conn[:,index2]]
        normalVector=np.cross(p1-p0,p2-p0)
        surfaceMetrics=np.linalg.norm(normalVector,axis=1) # small triangle

# integration in phi and r, using twice Lagrangian points approximation
for lp1 in np.arange(legPoints): # integration in phi
    for lp2 in np.arange(legPoints): # integration in r
        pLegendre = p0*zed[lp1,lp2]+p1*xi[lp1,lp2]+p2*eta[lp1,lp2]
        d = pLegendre - p0
        dd=np.einsum('ij,ik->ijk',d,d)
        r=np.linalg.norm(d,axis=1)[:,np.newaxis,np.newaxis]
        i = np.outer(np.ones(elNum),np.identity(3)).reshape(elNum,3,3)
        greenFunction = i/r + dd/r**3
        pre = np.pi/4*prefactorPhi[lp1]*radius[lp1,lp2]*legWeights[lp2]

        for el in np.arange(elNum):
        influenceMatrix[el::elNum,el::elNum] +=
    pre*surfaceMetrics[el]*greenFunction[el]

        influenceMatrix*= -1./(8.*np.pi*viscosity)

        return(influenceMatrix)
```

Where we used some new Numerical Python features. One very powerful one is the *Einstein Summation*, i.e. the possibility to perform multiplications of tensors of any rank just by repeating indexes. The instruction to do so is *np.einsum* and was essential to perform vectorized multiplication with the third rank Green Function T_{ijk}.

It is important to observe that in this implementation several for loops are employed, so it is not entire vectorized, however these loops are either through a fixed and small number of iteration ($6 \times 6 \times N_{LP} \times N_{LP}$) where N_{LP} is the number of Gauss Lagrange Points, and 6 stands for the number of nodes per element, therefore the overhead over the entire calculation of the matrix is minimal. Different would have been if I would have implemented two nested N_{el} loops, which would have scales like N_{el}^2. Still there are many margins of improvement and the reader is invited to explore them. This is also an excellent opportunity to implement a Cython compiled routine.

The result of the calculation of one of these Influence Matrices is shown in figure Fig. 11.3. Since every element of the mesh has a non-null interaction with every other triangle of the domain, the resulting Influence Matrix will always be dense. This clearly hampers the scalability of the this implementation for large problem. We already encountered dense matrices and understood it is extremely inefficient to employ them. When modelling boundary based problem the solution is the same as earlier, i.e. to devise an algorithm that allows solving the same problem without calculating these matrices at all. I will discuss later how this can be done, but let's first solve a problem with a full matrix, in order to understand how it works with its most straightforward implementation in Python.

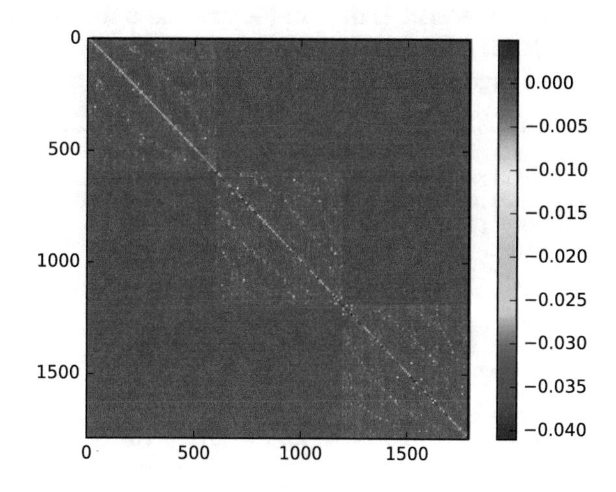

Fig. 11.3 Example of an influence matrix. The matrix is structured in three blocks referring to the x, y, z directions, related to the specific symmetry of the particle used for this example. The dominating terms are on the diagonal, due to the singular integrals. Only some terms out of the diagonal have non-negligible values. The presence of a large number of values close to zero implies that the matrix can be either conveniently compressed or that an alternative solving technique.

11.4.2 Calculation of the Resistance Matrix

Let's consider an ellipsoidal particle whose relative sizes in the x, y, z directions are 1, 1.5, 2 and let's assume that its equivalent Radius, defining its volume, is 1. These are just quantities that can be rescaled for the general case. We will calculate the drag over this particle exerted by a fluid with viscosity 1. Let's assume that *numpy* and *matplotlib.pyplot* are imported as *np* and *plt*. We have then:

```
radius=1.; ratiosBA=1.5.; ratiosCA=2.; equivalentRadius=1.
viscosity=1.

# create the mesh
nPoints=300
(allPoints,conn,elNum,pointsNumber) = createSecondOrderSphericalMesh(nPoints)
scale = equivalentRadius/(ratiosBA*ratiosCA)**(1./3.)
axis = np.array([scale,scale*ratiosBA,scale*ratiosCA])
allPoints[:,0:3]=axis[0:3]*allPoints[:,0:3]
```

where we used the function *createSecondOrderSphericalMesh()* introduced in Sect. 11.4 to create a second order triangular mesh of the ellipsoidal surface.

The next task is to find the position of the Gauss points for the integration. We will need the other function introduced in Sect. 11.4, *createGaussPositions()*. This function requires the predefinition of the Gauss Points for performing the non-singular integrations, and of the Legendre Points for performing the singular ones. For simplicity I predefine here 4 for the first and 6 for the second, as they guarantee an acceptable precision, however tabulated values exist everywhere in the literature with a large number of integration points, which guarantee great precision.

```
# integration parameters the singular integrals
legPoints=6; legCoords=np.zeros(legPoints); legWeights=np.zeros(legPoints)
legCoords[0] = -0.932469; legCoords[1] = -0.661209; legCoords[2] = -0.238619
legCoords[3:5] = -legCoords[2:0:-1]
legWeights[0] = 0.171324; legWeights[1] = 0.360761; legWeights[2] = 0.467914
legWeights[3:5] = legWeights[2:0:-1]

# integration parameters the non-singular integrals
gaussPoints=4;
xiCoord=np.zeros(gaussPoints);
etaCoord=np.zeros(gaussPoints);
gaussWeights=np.zeros(gaussPoints)
xiCoord[0] = 1./3.; xiCoord[1] = 1./5.; xiCoord[2] = 3./5.; xiCoord[3] = 1./5.
etaCoord[0] = 1./3; etaCoord[1] = 1./5; etaCoord[2] = 1./5; etaCoord[3] = 3./5
gaussWeights[0] = -27./48.; gaussWeights[1] = 25./48.
gaussWeights[2] = 25./48.; gaussWeights[3] = 25./48.

# create the collocation and gauss positions
initTime=time.time()
(gaussPositions, gaussMetrics) = createGaussPositions(allPoints,conn, elNum,
xiCoord, etaCoord, gaussPoints)
collocationPos = createGaussPositions(allPoints, conn, elNum, [1./3.], [1./3.],
  ↪ 1) [0]
[:,0,:]
```

Here one observes that *createGaussPositions()* was used twice. Once to calculate the positions of the Gauss Points, and the other to calculate the Collocation Position, i.e. the centroid of each Boundary Element.

We can now proceed and calculate the influence matrix, using the last function described above.

```
influenceMatrix=calculateInfluenceMatrix(elNum,
                gaussPoints, gaussWeights, gaussPositions,
                legPoints, legWeights, legCoords,
                collocationPos, gaussMetrics,
                allPoints, conn, viscosity)
```

The result of this calculation is shown in Fig. 11.3. The calculation of this dense matrix is clearly the most computationally demanding part of this implementation of the solver, as one can easily verify using the *%timeit* function of iPython.

Let's see now how to calculate a *Resistance Matrix* for the translational motion of the particles. In order to do so we have to create an array with the velocity in the directions x, y, z. Since the particle is rigid, the velocity will be equal for all the boundary elements, while of course the stress on every element will be different and depend on the complexity of the induced flow. Let's first create the right hand side, in the form of an array of the size of $N_{el} \times 3$ and solve the system of equations:

```
rightHandMatrix=np.zeros((3*elNum,3),float)
xyzInd1=np.arange(3)
xyzInd2 = np.arange(elNum*3).reshape(3,elNum).transpose()
deltaTr = np.ones((elNum,3)) # translational velocity
rightHandMatrix[xyzInd2[:],xyzInd1] = deltaTr

solution=np.linalg.solve(influenceMatrix,rightHandMatrix)
```

It is important to understand that we applied the solver function three times, as we solved for each of the x, y, z directions. Alternatively we could have solved only for the direction of motion of the particle, if we knew it. A standard technique to find the motion of rigid particles is to calculate the Resistance Matrix, and then solve for the velocity of motion of the particles. Clear for a system with many interacting particles the Resistance Matrix can be very complex and it might be necessary to iterate the solver with the contribution of all the particles.

To calculate the resistance matrix one needs to know also the area of each element, which I calculate here as half of the cross product between two sides of each triangle. It could be estimated even more precisely using the curvature properties of the function *createGaussPositions()*, but this is the most straightforward way.

```
elementsArea = 0.5*np.linalg.norm(
    np.cross(allPoints[conn[:,2]]-allPoints[conn[:,1]],
        allPoints[conn[:,1]]-allPoints[conn[:,0]]), axis=1).reshape(elNum,1)

prefactor = -1.0/(6*np.pi*equivalentRadius*viscosity)
resistanceMatrix=np.array(np.zeros((3,3),float))
```

```
for rightHand in np.arange(3):
    force=[0.,0.,0.]
    elementsStressNormal=np.array([solution[0:elNum,rightHand],
                    solution[elNum:2*elNum,rightHand],
                    solution[2*elNum:3*elNum,rightHand]]).transpose()
    force = np.sum(elementsStressNormal*elementsArea.reshape(elNum,1), axis=0)
    resistanceMatrix[0:3,rightHand] = prefactor*force[0:3]
```

The result of the calculation is shown in Fig. 11.4 for the motion in the z direction. The solution is technically the traction: $f_i(\mathbf{x}) = \sigma_{ik}(\mathbf{x})n_k(\mathbf{x})$ and the force on each face of the particle is the product of the traction for the area of the element.

For this precision one obtains a full Resistance Matrix similar to:

```
Translational Resistance Matrix

------------------------

 1.11181       7.50575e-06      -7.9172e-06
 6.6893e-06    1.02850          -5.675e-05
-1.0561e-05   -6.2623e-05        0.97108
```

Where one observes that the force is largely in the direction of motion of the particle. When investigating the drag over a particle, one should also calculate the rotational resistance, which can be done from the same influence matrix above, calculating the torque exerted against the rotation around one of the main axis.

This was just a quick introduction to a very flexible technique that has been applied to many fields of continuum mechanics. The method can be extended in

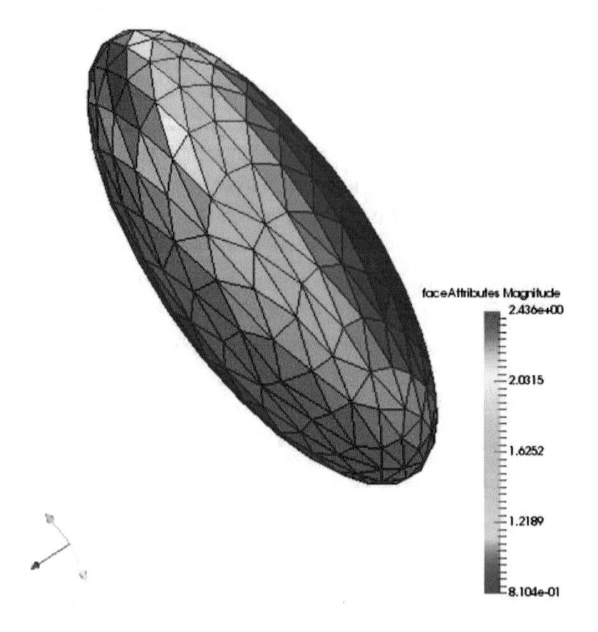

Fig. 11.4 Traction over the boundary elements of a rigid flat ellipsoidal particle calculated with the Boundary Element Method. Quadratic integration has been used on each element, performed with the Gauss Lagrange technique. The particle is here assumed to flow in the direction of its greater axis. The normal stress is larger where the curvature is greater

many directions. For example particles of any viscosity, also of zero viscosity, can be used, and therefore applied to modeling bubbles and multiphase flow in general. An application of this advanced technique is illustrated in Chap. 12.

The strategy described here is more of pedagogical use. For practical applications one can calculate the integrals using linear elements and analytical solutions, which allows some acceleration. However a greater gain is given by the combination of the tree-codes techniques described at the beginning of this chapter, which allows to totally avoid calculating the influence matrix, and invert the problem using instead a forward iterative method. This topic is quite advanced and the reader is invited to look at the scientific literature for updated techniques.

Summary

- Oct-trees offer a very simple space organization in which every cell has eight descendants equal in size and shape. The Barnes and Hut algorithm requires a maximum number of leaves in each terminal cell.
- Warren and Salmon envisaged a very fast and implementation of this algorithm using Hashing for every cell of the tree. If no more than one leave is allowed in the terminal cell, this allows building a tree from hashing only.
- Stokes Flow can be rewritten as a boundary equation. In particular a formulation due to Olga Ladyzhenskaya allows modeling the boundary between difference phases.
- Surfaces can be meshed in triangles and several integration techniques allow calculating the surface integrals of them. Main obstacles are the singularity that appears when a triangle interacts with itself and the N^2 scaling of the influence Matrix.
- Boundary Elements can be accelerated using Tree Codes, that replace entirely the calculation of the influence Matrix.

Problems

11.1 Write a vectorized version of the Hashing algorithm as illustrated in Sect. 11.1.2. The goal of creating a Hash Key ordering is to distribute the points in parallel environment in order to allocate nearby points in every processor. Can you reproduce a fast summation on 8 processors, accelerating its calculation by splitting the summation following the Hashing order?

11.2 Create a Boundary Element code that calculates the drag over a rigid particle using only linear triangles and calculate the integral using the analytical solution from [37]. Use both compressible and incompressible Green Functions and observe the difference.

11.3 Implement the boundary equation (11.6) using the same technique described in Sect. 11.3.1. Use either numerical integration for the *Double Layer* or an analytical formula from [37].

11.4 Use the Boundary Integral formulation from (11.1) to calculate the flow of the fluid around a particle. This integral does not display any singularity therefore it can be calculated with the simplest integration technique.

11.5 Advanced problem: combine the Boundary Element Method with the Warren and Salmon Tree strategy to create a fast BEM solver. Use the centroid to locate each element. Pay attention to that the resolution is sufficient to have no more than one element for every terminal cell of the Oct-tree. Use the distance between panels to establish a criteria on how to calculate the contribution of the single branches.

Chapter 12
Applications to Geodynamics

Abstract Several applications to Geodynamics are here synthetically addressed. Subduction, the driving mechanism of plate tectonics, can be simulated with all the techniques illustrated in this book. In particular by using the Particles in Cell (PIC) the details of the small scale thermo-mechanic processes can be modelled. Instead the Tree Code Boundary Element (BE) can have applications in modeling the far field interaction between tectonics plates. It is shown that the same technique can be applied to modeling the interaction between diapirs in magma and between crustal faults. Finally a traditional convection simulation is performed using the PIC method.

Sometimes the applications of the codes that we explored in this book is natural, other times it requires some further tweaks. I show in this chapter a few selected examples of applications where these strategies are applied. For more advanced applications and present challenges in the field of geodynamics there are many up-to-date reviews on the topics, e.g., [86].

Due to the limitation of space, I cannot write in this book the full implementation of the numerical codes behind the models that I show in this chapter. This review aims more at showing the potential of the numerical approaches that can be presently developed in Python, without using ad hoc geodynamic software.

12.1 Plate Tectonics

One of the most striking phenomena that has arisen during the evolution of the solid Earth is the tessellation of the surface in tectonics plates [3, 42, 113]. This phenomenon and its evolution in time is called *Plate Tectonics* and although an initial understanding of this process arose between the end of the sixties and the end of the seventies [147], still a vast number of geodynamic studies focuses on the understanding of the forces, the energetics and causes and the implications of this theory.

Open questions range from asking (i) when plate tectonics really started, (ii) whether in the Archean plate tectonics resembled as at present time, (iii) whether

G. Morra, *Pythonic Geodynamics*, Lecture Notes in Earth System Sciences,
DOI 10.1007/978-3-319-55682-6_12

is the mantle causing plates to move or are the plates driving the plate tectonics process [157], (iv) whether the upper and lower mantle of the Earth are coupled [56, 114, 133], (v) whether the *blobs* observed below the Pacific and the below Africa are involved in the general plate tectonics processes, (vi) whether mantle plumes are substantially fixed and therefore hot spots can be used to determine the past position of the tectonic plates [148], (vii) what is the real strength of the tectonic plates, and if elasticity plays a role [139, 145, 154, 160] (viii) why Earthquakes exist down to 670 km depth, and why not deeper, (ix) is the maximum magnitude of the thrust earthquakes related to the dynamics of the downgoing slab, and if so how, [136] (x) what is the energy budget of the plate tectonics phenomena, where is the energy thermally dissipated and whether it is generated internally in the mantle and flowing from the core into the mantle [143, 151, 158], (xi) whether the rheology of the mantle is strongly nonlinear, or if a linear behavior emerges at a certain scale [131, 150, 153], (xi) how are the global measured stress field and plate tectonics related [157], (xii) whether the evolution of plate tectonics on Earth is changing in a continuous manner or through sudden jumps of kinematic patterns [155, 159], (xiii) the relationship between the global geodynamics and the observed Earth's geoid [152] and many more questions. It is in summary an extremely active research field, just only looking at global and regional geodynamics open questions.

Classical plate tectonics theory has achieved many successes at explaining many features of regional kinematics, i.e., how a single tectonic plate behaves [146, 149]. The role of the larger scale, of how plates interact with each other in a global setting [162], and how new tectonic plates are created and destroyed, is still very open [161]. The same can be said for the lower mantle, which still holds a large number of mysteries, among them (i) the role of the post-perovskite, a last phase of the mantle that exists only near the bottom of the lower mantle, in proximity of the core, (ii) the role of the spin transition that characterized Fe bearing minerals in the lower mantle [138], and again many more.

All the techniques that I have introduced in this book have been used for modeling regional plate tectonics (subduction, and ridge dynamics, mainly), as well as global plate tectonics. For example, the Boundary Element Method described in the past chapter has been used to model the dynamics of a diverging ridge. For a lithosphere model with 100 times the viscosity of the mantle, the characteristic ridge topography arises self-consistently in the surface of the Earth as depicted in Fig. 12.1. The dynamics begin with a transient regime where the lithosphere sinks and the surface of the Earth adapts until the system relaxes to a state of equilibrium, to a morphology composed by a seamount diluting in the far field to a flat plateau. In this stable state topographic and mechanical features can be compared with known data.

To understand the role of elastic stresses, the simulation is performed in two stages. Initially a Stokes Flow simulation is run until equilibrium is reached, then for that configuration elastic integrals are calculated. The elastic stresses are described with the same formulation of the Stokes flow, if the strains replace strain rates. However the Boundary Equations usually change because the material requires different boundary conditions. The BC imposed can represent every additional force that modifies the state of stress on the plate. The displacement field is sampled in a 3-dimensional

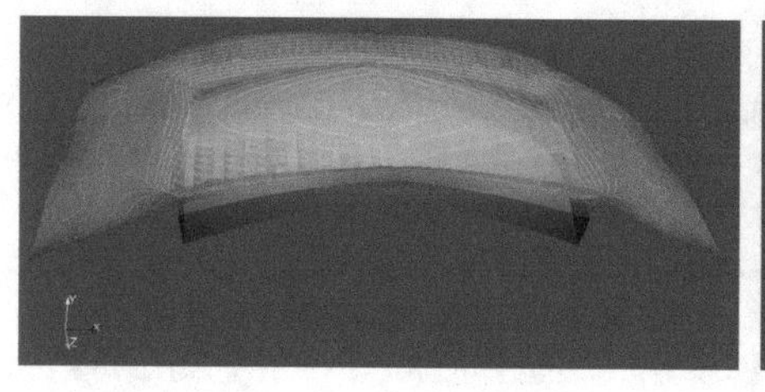
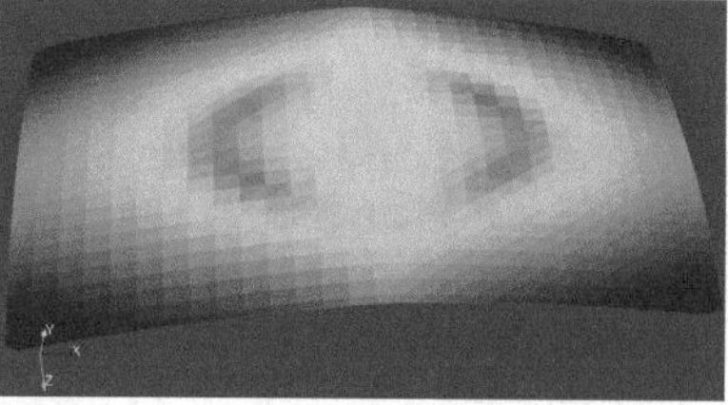

Fig. 12.1 *Left* Mid ocean ridge topography emerging from dynamical evolution of lithosphere–mantle–surface interaction. *Right* Second stress invariant on the lithosphere surface for the same simulation

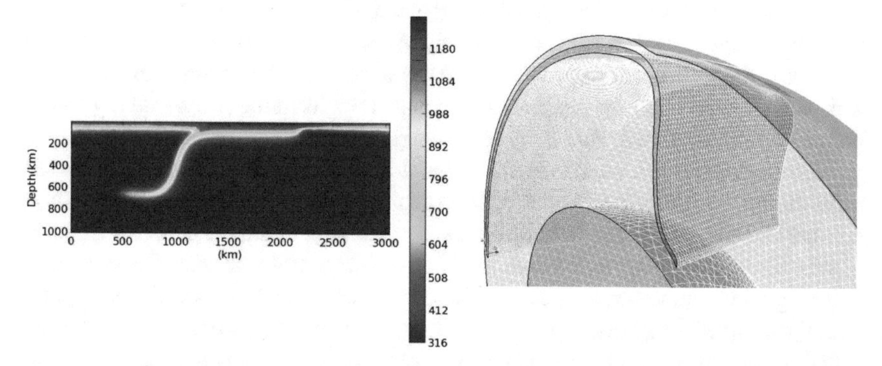

Fig. 12.2 *Left* Thermo-mechanical model (temperature is visualized) of a subducted slab modelled with the Particles in Cell Method. The particles allow precise tracking of the slab morphology while the background lattice allows calculating the solution of the Momentum and Energy equations. *Right* Model of a developed downgoing plate at the global scale modelled using exclusively the Boundary Element Method described in Chap. 11

region of interest and used to construct a stress map which is shown in Fig. 12.1. Though buoyancy holds the lithosphere in isostasy there is a boundary effect on the edges of the plate. The simulation is taken from [156].

The same technique has been used for modeling subduction at the regional and global scale, as shown in Fig. 12.2.

12.2 Raise of Gas in a Volcanic Conduit

Magmas are multiphase flows transporting solid and gas components that undergo intense bulk and shear deformation [92, 119]. The calculation of the average rheology of a complex fluid is in general a very challenging task, and in the case of magma is an even harder problem due to the impossibility to take direct measurement in an active volcanic setting, given that no material can withstand the temperatures of molten magma. For these reasons all the data that models can match, besides historical and geological reconstructions, are from seismic, acoustic and gravity measurements.

The gas component is always involved in volcanic eruptions. In fact the three main types of eruptions are (i) magmatic (magma+gas, but magma driven), (ii) phreatomagmatic (magma+gas, but gas driven), (iii) phreatic eruption (steam superheating due to contact with magma). While the classification of the large variety of volcanic activity, from slow eruption to explosive [117], illustrates how the system dynamics can diverge, many volcanic conduit release their gas and magma through a frequent and repetitive controlled explosive manifestation called *Strombolian Activity* [118].

Many works have been devoted at understanding how bubbly dynamics from Strombolian activity [121]. In such type of activity, the gas component inside the ascending magma is separated in bubbles that expand with the rise of the magma and the consequent reduction of pressure [116]. As a consequence the bubbles expand, increasing the speed of their dynamics and the amount of interaction with each other. This work focuses on how this interaction acts, how its numerical modeling can be numerically tackled and what early results suggest [122].

Many advances to understand the dynamics of this system can be obtained by using Boundary Element of Stokes flow to capture the collective dynamics of a large number of Bubbles. In the limit where the deformation of the bubbles is negligible, when the surface tension is sufficiently strong to compensate the shear forces applied to a single bubble, the dynamics of a large number of bubbles can be investigated with the simpler Singularity Method, using the fundamental Stokes' flow solutions called Stokeslets as recently employed in geophysical fluid dynamics [126] and biofluid dynamics [127]. It the Strombolian Activity is a manifestation of the nonlinear interaction of the rising bubbles, which is still topic of research [125], it would another geophysical phenomena analogue to many other complex systems [120]. Models shown in Fig. 12.3 for example, illustrate how many bubbles, even at distance, will interact through their induced 3D flow and create complex dynamics, often impossible to predict without accurate modeling (or laboratory experiments).

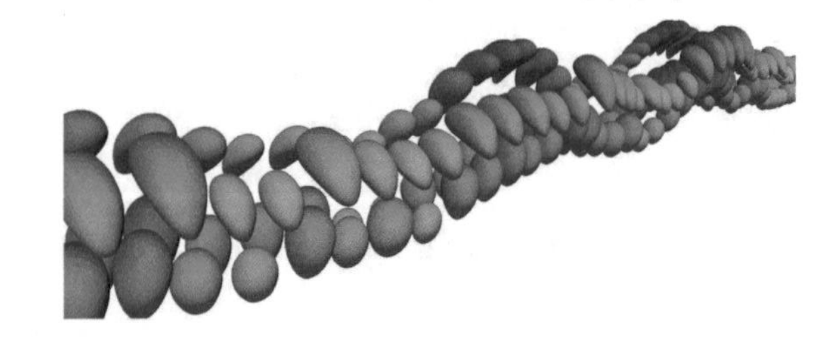

Fig. 12.3 Model of the rise of a long channel of bubbles in a viscous fluid. Although the rheology of the fluid is linear, the close and far field interaction between the sphere generates a complex large pattern, initially undulatory, as in this figure, and then creating localized plumes of bubbles. The technique employed here is the Tree Code Boundary Element Method described in Chap. 11

12.3 Interaction Between Faults

One of the main advantages of using the Boundary Element Method is that it allows modeling faults and their interaction at any scale. For example in Fig. 12.4 I show the flow induced into two displaced normal faults. The advantage of using the BEM is clear since only the elements representing the faults is required.

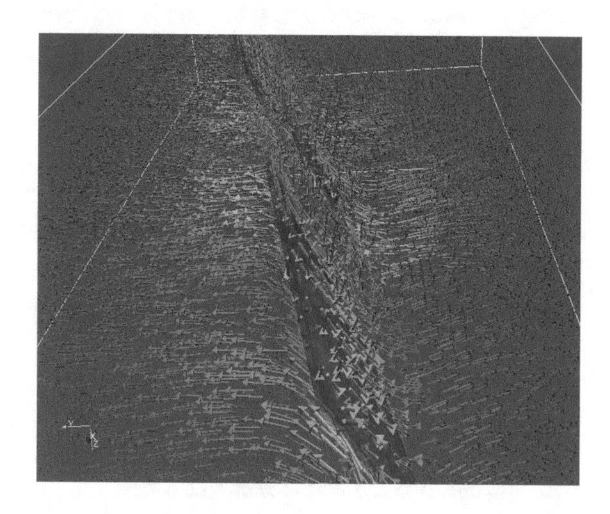

Fig. 12.4 Model of a two faults, one displaced from the other, modeled by the Boundary Element Method described in Chap. 11

12.4 Convection in 2D

The Stokes solution and the diffusion solution can be combined to solve the traditional convection problem, i.e., the flow of a highly viscous fluid when heated from below. Assuming constant thermal expansion, the fluid will expand linearly with the temperature and will tend to rise due to buoyancy. This forcing is, however, compensated by the heat diffusion. If the diffusion is sufficiently powerful it will disperse the heat before the instability (in general in the form of a plume) can rise. The simplest solution to this stability problem is condensed in the Rayleigh number. The Rayleigh number for bottom heated very wide layer is:

$$Ra = \frac{\rho_0^2 g \beta \Delta T D^3 C_P}{\eta k} \tag{12.1}$$

where ρ_0 is the reference density at the temperature of the upper plate T_0, D is the depth of the fluid layer, η is the viscosity, ΔT is the temperature difference between the bottom and upper plate, k is the thermal conductivity, and C_P is the specific heat capacity. This expression does not contain pressure, which implies that for the mantle ΔT is the superadiabatic temperature (difference between the real temperature and the one increased by the compression alone).

The above expression allows calculating the stability for Earth in the simplified assumption of constant viscosity, and the verification of the critical number using the streamline approach is an interesting test. In particular to calculate the time needed for the onset of convection versus the Rayleigh number. However as soon as the very strong temperature dependency of the temperature is added, to calculate the onset of convection numerical methods become necessary. Let us, therefore, see how the nonlinear Stokes solver can be combined with the upwind advection scheme to obtain a simple convection code. The procedure is very linear:

(A) Start with an initial linear temperature gradient plus random initial perturbations
(B) Calculate the solution of the Stokes Equation
(C) Advect the temperature (Upwind scheme)
(D) Diffuse the temperature (implicit)
(E) Apply the Boundary conditions for Temperature at the top and the bottom

Maxwell–Boltzmann statistical mechanics tell us that the most probably shape of a temperature dependent viscosity has the shape:

$$\eta(T) = \eta_0 e^{-\frac{E}{RT}} \tag{12.2}$$

where E is the activation energy and $R = kN_a$ is the gas constant, the product of the Boltzmann constant k for the Avogadro number N_a. Imposing this law to a standard convection equation, and calculating only the deviation of the temperature from the adiabatic one, one can write the simple code:

```
LaplacianOp=sparseLaplacianOperator(nxp,nyp,dx,dy)
LaplacianOp=addBC(LaplacianOp,nxp,nyp) coeff =
np.log(viscContrastTemp) for step in np.arange(steps):
    viscosity[:,:] = np.exp( -coeff * (T[:,:]-0.5) )

    # Right Hand Side (= g*rho = -Ra*T)
    rhs[0:nxp,1:nyp] = -Ra * (T[0:nxp,0:nyp-1] + T[0:nxp,1:nyp]) / 2.0

    # solve for velocity and pressure
    fullStokesOp = fullStokesOperator(viscosity,dx,dy,nxp,nyp)
        solution = la.spsolve(fullStokesOperator,rhs)

        vx=solution[0:nxy].reshape(nxp,nyp); print(vx.shape)
        vy=solution[nxy:2*nxy].reshape(nxp,nyp)
        p=solution[2*nxy:3*nxy].reshape(nxp,nyp)

    # determine stable time-step
    absVx=np.abs(vx); absVy=np.abs(vy);
    vMax = np.max([absVx.max(),absVy.max()])
    dt = 0.5 * np.min(dx,dy) / vMax

    # Diffusion
    IA=sparse.eye(nxp*nyp).tocsc()-LaplacianOp*r*dx*dx  # I-A
    T = la.spsolve(IA,T.reshape(nxp*nyp)).reshape(nxp,nyp)

    # Advection of the temperature
    T = vectorizedUpwind(T,vx,vy,dx,dy,dt)

    # Set top box temperature to zero:
    T[:,0] = 0.
```

The results of this simulation where *viscContrastTemp* is 10^6 is shown in Fig. 12.5.

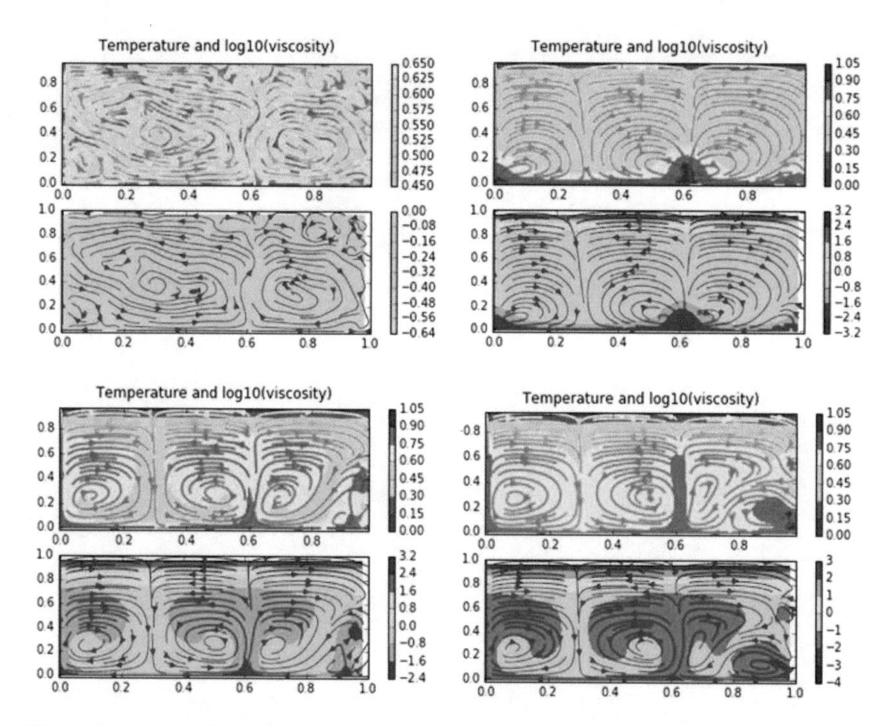

Fig. 12.5 Four snapshots of a convection simulation with temperature controlled by a Boltzmann term and a given Rayleigh number. The temperature at the *top* of the box is fixed. This type of convection is paradigmatic for the simulation of mantle convection. Recently scientists have added phase transitions and plates can explain features not present in this model.

Chapter 13
The Future

"If you think you are too small to make a difference, try sleeping with a mosquito."

— Dalai Lama XIV

Abstract I outline a possible future for Python and the Geosciences. We are now immersed in a coming revolution in which new artificial intelligence tools, mainly deep neural networks, are emerging and seem to be able to revolutionize geosciences by offering new ways to analyze data. We cannot predict the future applications of these tools, but we can foresee that the tools explained in this book will remain important for at least another decade and that easy to learn and to use but powerful languages like Python will play a central role.
When should students and professionals who will work in the field of geosciences learn computational methods? At present in most colleges in the world, computing, and in particular high-performance computing, is postponed to the last years of a graduate course, or directly to postgraduate studies. This will inevitably change in the next future because to understand computers, how these machines work, what they can do for us and how we can use them to improve the quality of our work and research is essential, and the sooner students learn fundamental computing techniques, the better. In this final chapter, I try to describe in very general terms some of the key ingredients that in my view will picture the future of computational geodynamics. I am aware that some key innovations is missing from this list, however I am certain that the following ones will become essential for every computational scientist, and geoscientist in particular.

13.1 Jupyter

In Sect. 2.4, I have mentioned how iPython has evolved into Jupyter Notebooks. I would like to emphasize how this new way to transmit scientific content is potentially revolutionary. The main tools that we have today to communicate scientific findings

G. Morra, *Pythonic Geodynamics*, Lecture Notes in Earth System Sciences,
DOI 10.1007/978-3-319-55682-6_13

are publications and books. These two have been working perfectly for centuries, however since today our main means of collecting and transmitting information is the Internet, we are in need of a new tool. Or better, because computers and the Internet are so flexible, we will probably have many tools. More specifically referring to numerical modeling, the main obstacle in communicating results is the lack of detailed descriptions of the algorithms themselves. This volume tries to progress toward a *hybrid* between a book and a pure software. While writing, however, I found myself at uneasy with the written page, and I believe that in the future we will communicate techniques and numerical strategies with Notebooks, where the details are all spelled out, that can be modified and played with, as it is the spirit of this volume.

Notebooks in iPython and in Jupyter allow to express and *visualize* our reasoning and scientific argument in a precise manner. If we look retrospectively at the advancements of science, we see that most progresses have not happened because of the brilliant intuition of a great mind, but most often because of the hard work of developing a new technique/technology that has allowed a discovery. Unfortunately scientific papers are often opaque in their capability to share technical features. For this reason Jupyter Notebooks, or a new technique that may be even more precise and effective, will become the key to communicate computational scientific applications in a new, fresh and modern way.

One recent example of how to use Notebooks in geoscience is EarthPy (`earthpy.org`). This is a collection of IPython notebooks related to Earth Science. Most of the Notebooks are tutorials that combine a short description with an in-depth example showing the key tuning and trick necessary to make a Python pipeline to work. Several of the Notebooks aim at a better visualization, not only of maps, but also of statistical data. Certainly this project and similar ones will continue to develop in the future making the distribution of Earth Python resources more efficient.

13.2 Machine Learning

We live in extraordinary times. It was well 20 years ago when the world chess champion Garry Kasparov lost at tournament conditions against Deep Blue, an intelligent machine built by IBM. That computer was programmed with the same style that I have been teaching in this book. Have an idea, implementing smart algorithms, debugging, testing.

After chess, only one table game was left in which humans were completely superior. That was Go, a game so complex that its theoretical organization was never formalized. Like when an expert watches painting or a mountain guide observes the landscape on a trail, Go top players are driven by intuition and have been trained by sitting in front of a greater master and learned by practicing the game. This year the surprising but neat victory of Alpha Go, a computer powered by deep learning tools and self-trained only few months, against the world reigning Go champion, has probably opened a new page in the history of the human-machine interaction.

It is possible that Alpha Go has invented a new Go playing style, or that it has learned to "feel" the next good move. We do not know yet. Still as numerical modelers we must acknowledge that we are at a turning point. If computers can now efficiently mimic the human cognitive system is such a powerful way to be able to reach hyper-human performance in several fields, in geodynamics, and more in general in computational geosciences, the future will be to use deep learning tools not only to interpret data but even more to create numerical models. In fact we can think at playing chess, autonomous driving, and more as extraordinary numerical models.

Today artificial intelligence tools can recognize which animal is on a picture. Even the breed of a dog, which is a feature superior to many humans. Looking back at the (limited number of) numerical tools that I have introduced in this book, one cannot avoid to observe that much of the work of the developer can be made automatically. For example, parameters tuning when building matrix blocks. Much of the debugging process can be done also automatically. And even solving clearly defined physics problems will be soon possible for intelligent systems.

As a practical example, I have described in this book some sophisticated and (presently) advanced techniques to collect and organize a large number of points in a tree in order to reduce the number of operations (from N^2 to $Nlog(N)$) necessary to calculate the overall interactive force. While these algorithms are fixed and devoted to solve some specific problem like the gravity force of a million of bodies in the space, the interaction of thousands of rigid particles moving in a viscous fluid, or the overall flow generated by a billion of vortex particles in turbulent flow, AI powered system promise to deliver a generic creator of algorithm for fast many-body calculations. This is possible because these tree codes are at their core just a subfield of network sciences.

It is impossible to see now the speed at which this evolves, but we can see the direction, which is to offer to scientists new tools that will disrupt old techniques as much as the introduction of personal computers has done 30 years ago. With open systems like Tensor Flow (`https://www.tensorflow.org/`), *Theano* (`http://deeplearning.net/software/theano/`), and Torch (`http://torch.ch/`) all of them freely open to developers, we can be sure that the this field will evolve quickly. The importance of learning Python today is in fact made clear by the languages in which these tools are made available. The APIs of Tensor Flow, the major product of the Brain Google team, are available only in C++ and Python. Theano is only available in Python, while Torch in *LuaJIT*, that is in many ways analogue to Numerical Python, being specifically devoted to run on GPUs. We can predict that within a decade scientists will solve the problems that I have introduced in this book using this new technology to generalize our algorithms to a broader range of problems (e.g., natural implementation of plasticity, general tree codes with any possible kernel).

13.2.1 Theano and Tensor Flow

Theano is an academic project aimed at building a Python based deep-learning library with all the necessary tools necessary to work with multilayer Neural Networks. Practically Theano is the predecessor of Tensor Flow (TF), which is a project developed by Google with the aim of producing a much more powerful machine learning.

Tensor Flow has been built not as a general AI or Machine Learning tool, but more as an extension analogue to Numpy. It is structured similarly as a library of functions that apply on n-dimensional arrays. The main difference is that NumPy does not offer built-in methods to create tensor functions and calculate the derivatives that most Machine Learning tools need. TensorFlow instead provides the necessary primitives for defining functions on tensors and their derivatives.

It is important to notice that tensors in TF are different from tensors in Physics. In TF, tensors are simply two-dimensional arrays. In Physics, instead tensors have specific physical properties by which they are defined in function of the physical space and define invariants respect to operations like rotation and displacement. For example the definition of a linear momentum in physics, that is a vector, does not depend on the numbers used, which can depend on the definition of the x, y, and z axis. Similarly, the numbers used to describe the Stress Tensor in a certain point for a deforming body can change by changing reference frame, but the tensorial physical quantity does not change.

In TF, the word tensor is used in a more general framework to indicate a two-dimensional array. Probably the most important difference between NumPy and Tensor Flow is that TF does not immediately calculate the result of an operation, defining instead a *Computation Graph*. For example, if A and B are two tensors, their product $C = A \times B$ in TF is not the the computation of the matrix product, but only a *node* in a *Graph*. In order to establish the context where the *Graph* can be run, in TF one defines a *Session* and only the *Graph* defined in the *Session* will be computed. In practice TensorFlow programs are organized in a (i) construction phase, in which the *Graph* is assembled and (ii) a session in which the operations in the *Graph* are executed.

Let us see an example:

```
# create and multiply matrices with tensorflow
import tensorflow as tf

n = 1000
a = tf.random_uniform((n,n))
b = tf.random_uniform((n,n))
c = tf.matmul(a,b)

with tf.Session() as sess:
    sess.run(c)
    print(c.eval())
```

Where the product between a and b was computed only when the command *sess.run(c)* was given. Clearly this was a simple example, but the reason for developing a *Computation Graph* is that for much larger problems the *Graph* can be visualized, modified, optimized, and so on. This is also a simple example because it involves only constant arrays (a and b are constant once they have been defined). By considering large complex graphs that involve variables, the benefits of using this approach emerge more clearly.

Besides *Computation Graph* another property makes tools like Theano and Tensor Flow unique, which is its ability to optimize complex problems without explicitly defining the solution technique. How this is performed is shown in the following example, where I first create a cloud of points around a line in two dimensions, and then I ask Tensor Flow to find the linear fit ($y = ax + b$) that fits the cloud, where the optimal fit is defined as the one that minimizes the sum of the error square. This is called linear regression and is the simplest and most used optimization strategy. Normally this is obtained by explicitly defining the formula the minimizes the errors, which consists in calculating the gradients and using them to find increments toward the solution, which is approximated after a certain number of increments.

Let us see how this is automatically calculated in Tensor Flow by calculating a very simple linear regression fit. Generally, one manipulates the given data in NumPy. So I create first a set of linear data and then perturb it with random noise:

```
import tensorflow as tf
import numpy as np
import matplotlib.pyplot as plt

# Given parameters
aGiven = 0.5; bGiven = 5.0; noise = 0.1

# Given Data
nSamples = 100
trainX = np.arange(nSamples)*1./nSamples
trainY = trainX * aGiven + bGiven
trainY += noise * np.random.randn(nSamples)
```

Machine Learning is normally used with large sets of data, so it is generally not practical to create a copy of all the data to be accessed by the Computational Graph. In Tensor Flow, the solution is to create *place holders* for the given data. Once this is done, it is necessary to set the learning parameters (*Learning Rate*, *Training Steps*) of the model that we believe a priory fits the system better (linear in our case) and the *cost function* that we want to minimize.

```
# Tensor Flow Graph Input
X = tf.placeholder('float')
Y = tf.placeholder('float')

# Learning Parameters
learningRate = 1.0
maxTrainingSteps = 100
```

```
# Start with some random values
aLearned = tf.Variable(np.random.randn(), name='first')
bLearned = tf.Variable(np.random.randn(), name='second')

# Construct a linear model
model = tf.add(tf.mul(X, aLearned), bLearned)

# Cost Function: mean squared error
cost = tf.reduce_sum(tf.pow(model-Y, 2))/(2*nSamples)
```

We use here the Gradient Descent minimization algorithm, that is just the analogue in many dimensions of the Newton method for finding the zeros of an equation, and is the most commonly used technique in Machine Learning. Again as before we do not have to instruct TF on how to calculate the gradients. This is really the strong advantage of using TF. After that one can just iterate within a TF session and compared the learned fit vs the given one. I do this in figure Fig. 13.1.

```
# Gradient descent
optimizer = tf.train.GradientDescentOptimizer(learningRate).minimize(cost)

# Initializing the variables
init = tf.global_variables_initializer()

# Launch the graph
with tf.Session() as sess:
    sess.run(init)
    c = 1.0; step = 0
    # Fit all training data
    for step in np.arange(maxTrainingSteps):
        for (x, y) in zip(trainX, trainY):
            sess.run(optimizer, feed_dict={X: x, Y: y})

        c = sess.run(cost, feed_dict={X: trainX, Y:trainY})
        print('Step:', step, 'cost=', c, \
              'learned a=', sess.run(aLearned), \
              'and learned b=', sess.run(bLearned))
```

This example was extremely simple and a NumPy implementation of the same algorithm for this specific dataset would be much faster, however the power of TF is that it can be generalized to large projects with hundreds of variables without the need to figuring out how to optimize the problem. This implies that for very complex problems with many minimums, as it is often the case in recognition and in post-processing modeling data, this approach can be extremely powerful. The deep understanding and the many technical details necessary to develop a working model that can be applied to solving PDEs, or to organize model outcomes, or to organize geophysical data goes well beyond the scopes of this book and can be learned by many sources such as [18, 19], and of course on `https://www.tensorflow.org/`.

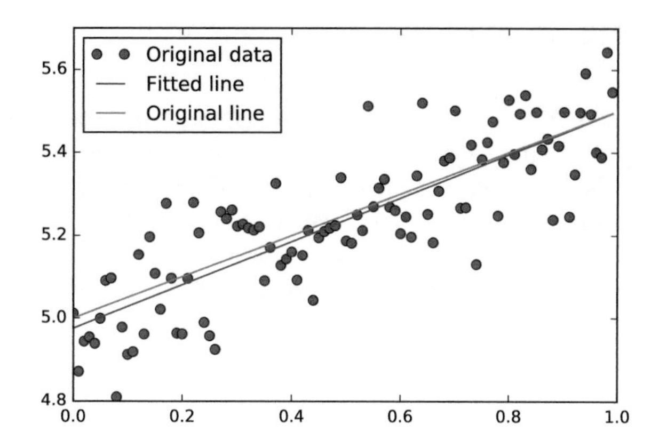

Fig. 13.1 In *green* the *original line*. *Red dots* are extracted from the given *line* plus a random perturbation. The machine learning algorithm finds the *fitted line* after hundred steps by minimizing the cost function given by the sum of the error squared. No gradient to follow is explicitly given to Tensor Flow

13.3 Big Data

There is another key aspect that is more specific to geophysics, or better to Geosciences as a whole. Earth studies are about earth's interiors, oceans, biosphere, atmosphere, magnetosphere, and so on, so the interpretation of the mass of data that originates from our Planet as a whole promises to revolutionize the entire field. Just to make few examples, today real-time waveforms from seismic events are immediately available online from sites such as `http://earthquake.usgs.gov/monitoring/helicorders.php`. We can envisage a near future in which automatic systems will scan the Internet and extract the data in whatever format they are and make them immediately available to scientists. Another clear revolution will be the elaboration of data from gliders, Underwater Autonomous Systems and others, capable of collecting immense high-resolution images, acoustic data and more. They will need a completely new generation of post-processing tools that will combine the easiness to use that Python has, with a far greater power and flexibility that similar languages will have.

It is important to point out that Big Data and artificial intelligence, although certainly revolutionary for the next future of scientific computing, are based on very different algorithms compared to how humans learn. Children do not learn from the detailed processing of huge datasets, but they learn by observing few small cases from which they can extrapolate general laws. And this is also how scientific laws have been derived. This implies that Big Data interpretation, and the fundamental role that Python and other high-level and high-performance languages will have in this development, will only be complementary to the ability to make choices and finding general rules that at the present only humans seem to be able to do.

13.4 Final Outlook

Overall high-level languages like Python promise to have a central rule in the future of geodynamics and geosciences in general. These innovations have also the potential to make geosciences more ethical. If it is incontestable that we as humans are changing our environment at every level, I see as the ultimate goal of geoscientists to monitor the *health* of our Planet. In this sense I see the future of numerical modeling to be an essential tool of the much broader Big Geo-Data revolution.

References

1. ABAQUS/Standard (2000). Theory Manual, Hibbit, Karlsson and Sorenson Inc.
2. Bengio, Y., Goodfellow, I. J., & Courville, A. (2015). Deep learning. http://www.deeplearningbook.org
3. Bird, P. (2003). An updated digital model of plate boundaries. *Geochem. Geophys. Geosyst.*, *4*(3), 1027. doi:10.1029/C000252.
4. Bodenheimer, P., Laughlin, G. P., Rozyczka, M., Plewa, T., & Yorke, H. W. (2006). *Numerical methods in astrophysics: An introduction*. London: Taylor & Francis.
5. Brebbia, C. A., & Walker, S. (2016). *Boundary element techniques in engineering*. Amsterdam: Elsevier.
6. Chen, Z., Huan, G., & Ma, Y. (2006). *Computational methods for multiphase flows in porous media* (Vol. 6). Philadelphia: Siam.
7. Chityala, R., & Pudipeddi, S. (2014). *Image processing and acquisition using Python*. Boca Raton: CRC Press.
8. Cottet, G. H., & Koumoutsakos, P. D. (2000). *Vortex methods: theory and practice*. Cambridge: Cambridge University Press.
9. Courant, R., & Hilbert, D. (1965). *Methods of mathematical physics* (Vol. 1). CUP Archive.
10. Davies, G. F. (1999). *Dynamic Earth: plates, plumes and mantle convection*. Cambridge: Cambridge University Press.
11. Gerya, T. (2009). *Introduction to numerical geodynamic modelling*. Cambridge: Cambridge University Press.
12. Guazzelli, E., & Morris, J. F. (2011). *A physical introduction to suspension dynamics* (Vol. 45). Cambridge: Cambridge University Press.
13. Guttag, J. V. (2013). *Introduction to computation and programming using Python*. Cambridge: MIT Press.
14. Halliday, D., Walker, J., & Resnick, R. (2014). *Fundamentals of physics*. London: Wiley.
15. Happel, J., & Brenner, H. (1983). *Low reynolds number hydrodynamics*. The Hague: Martinus Nijhoff.
16. Hockney, R. W., & Eastwood, J. W. (1988). *Computer simulations using particles*. Bristol: Institute of Physics Publishing.
17. Idris, I. (2015). *NumPy Cookbook*. Birmingham: Packt Publishing Ltd.
18. Ivezic, Z., Connolly, A. J., VanderPlas, J. T., & Gray, A. (2014). *Statistics, data mining, and machine learning in astronomy: A practical Python guide for the analysis of survey data*. Princeton: Princeton University Press.
19. Kelleher, J. D., Mac Namee, B., & D'Arcy, A. (2015). *Fundamentals of machine learning for predictive data analytics: Algorithms, worked examples, and case studies*. Cambridge: MIT Press.

G. Morra, *Pythonic Geodynamics*, Lecture Notes in Earth System Sciences,
DOI 10.1007/978-3-319-55682-6

20. Kennett, B. L. N., & Bunge, H. P. (2008). *Geophysical continua: Deformation in the Earth's interior*. Cambridge: Cambridge University Press. ISBN 978-0-521-86553-1.
21. Kim, S., & Karrila, S. J. (2013). *Microhydrodynamics: Principles and selected applications*. Mineola: Courier Corporation.
22. Koschmieder, E. L. (1993). *Benard cells and Taylor vortices*. Cambridge: Cambridge University Press.
23. LeVeque R (2007). Finite Difference Methods for Ordinary and Partial Differential Equations: Steady State and Time Dependent Problems, Society for Industrial and Applied Mathematics (SIAM), Philadelphia. ISBN 978-0-898716-29-0
24. Lowrie, W. (2011). *A student's guide to geophysical equations*. Cambridge: Cambridge University Press.
25. McKinney, W. (2012). *Python for data analysis*. Newton: O'Reilly.
26. Minsky, M. (1988). *Society of mind*. New York: Simon and Schuster.
27. Morra, G., Yuen, D. A., King, S. D., Lee, S. M., & Stein, S. (2015). *Subduction dynamics: From mantle flow to mega disasters*. London: Wiley.
28. Ogawa, M., Schubert, G., & Zebib, A. (1991). Numerical simulations of three-dimensional thermal convection in a fluid with strongly temperature-dependent viscosity. *Journal of Fluid Mechanics*, *233*, 299–328.
29. Pacheco, P. S. (1997). *Parallel programming with MPI*. San Francisco: Morgan Kaufmann.
30. Patankar, S. V., & Spalding, D. B. (1972). A calculation procedure for heat, mass and momentum transfer in three-dimensional parabolic flows. *International journal of heat and mass transfer*, *15*(10), 1787–1806.
31. Patankar, S. (1980). *Numerical heat transfer and fluid flow*. Chicago: CRC Press.
32. Pozrikidis, C. (1992). *Boundary integral and singularity methods for linearized viscous flow*. Cambridge: Cambridge University Press.
33. Pozrikidis, C. (2002). *Boundary element method*. Boca Raton: Chapman and Hall/CRC.
34. Pozrikidis, C. (2011). *Introduction to theoretical and computational fluid dynamics* (2nd ed.). Oxford: Oxford University Press.
35. Prosperetti, A., & Tryggvason, G. (Eds.). (2009). Computational methods for multiphase flow. Cambridge university press.
36. Ranalli, G. (1995). *Rheology of the earth*. London: Chapman and Hall.
37. Salvadori, A. (2010). Analytical integrations in 3D BEM for elliptic problems: evaluation and implementation. *International Journal for Numerical Methods in Engineering*, *84*(5), 505–542.
38. Scopatz, A., & Huff, K. D. (2015). *Effective computation in physics*. Newton: O'Reilly Media Inc.
39. Shearer, P. M. (2009). *Introduction to seismology*. Cambridge: Cambridge University Press.
40. Smith, K. W. (2015). *Cython - A guide for Python programmers*. Newton: O'Reilly.
41. Snir, M., & Gropp, W. D. (1998). *MPI? the complete reference*. Cambridge: MIT Press.
42. Sornette, D., & Pisarenko, V. (2003). Fractal plate tectonics. *Geophys. Res. Lett.*, *30*(3), 1105. doi:10.1029/2002GL015043.
43. Stauffer, D., & Stanley, H. E. (1996). *From newton to mandelbrot: A primer in theoretical physics with Fractals for the Macintosh* (2nd ed.). Berlin: Springer.
44. Tao, T. (2012). *Topics in random matrix theory. Graduate studies in mathematics* (Vol. 132). American Mathematical Society, Providence, RI. MR-2906465
45. Torquato, S. (2013). *Random heterogeneous materials: Microstructure and macroscopic properties*. Berlin: Springer.
46. Turcotte, D., & Schubert, G. (2002). *Geodynamics*. New York: Cambridge University Press.
47. Anderson, J. D., & Wendt, J. (1995). *Computational fluid dynamics* (Vol. 206). New York: McGraw-Hill.
48. Ferziger, J. H., & Peric, M. (2012). *Computational methods for fluid dynamics*. Berlin: Springer Science and Business Media.
49. Colella, P., & Puckett, E. G. (1998). Modern numerical methods for fluid flow. University of California. Lecture notes available at http://www.amath.unc.edu/Faculty/minion/class/puckett/C_P_Notes.pdf.

50. LeVeque, R. J. (2002). *Finite volume methods for hyperbolic problems (Vol. 31)*. Cambridge: Cambridge university press.
51. Ladyzhenskaya, O. A. (1969). *The mathematical theory of viscous incompressible flow* (Vol. 76). New York: Gordon and Breach.
52. Slingerland, R., & Kump, L. (2011). *Mathematical modeling of Earth's dynamical systems: A primer*. Princeton: Princeton University Press.
53. Balay, S., Abhyankar, S., Adams, M. F., Brown, J., Brune, P., Buschelman, K., Dalcin, L., Eijkhout, V., Gropp, W. D., Karpeyev, D., Kaushik, D., Knepley, M. G., Curfman McInnes, L., Rupp, K., Smith, B. F., Zampini, S., Zhang, H., & Zhang, H. PETSc Users Manual, ANL-95/11
54. Braun, J., & Sambridge, M. (1995). A numerical method for solving partial differential equations on highly irregular evolving grids. *Nature*, *376*, 655–660.
55. Bell, J. B., Colella, P., & Glaz, H. M. (1989). A second-order projection method for the incompressible Navier-Stokes equations. *Journal of Computational Physics*, *85*(2), 257–283.
56. Christensen, U. (1984). Convection with pressure-and temperature-dependent non-Newtonian rheology. *Geophysical Journal International*, *77*(2), 343–384.
57. Dabrowski, M., Krotkiewski, M., & Schmid, D. W. (2008). MILAMIN: MATLAB?based finite element method solver for large problems. *Geochemistry, Geophysics, Geosystems*, *9*(4).
58. Dalcin, L. D., Paz, R. R., Kler, P. A., & Cosimo, A. (2011). Parallel distributed computing using python. *Advances in Water Resources*, *34*(9), 1124–1139.
59. Goldberg, D. (1991). What every computer scientist should know about floating-point arithmetic. *ACM Computing Surveys (CSUR)*, *23*(1), 5–48.
60. Quarteroni, A., Sacco, R., & Saleri, F. (2010). *Numerical mathematics* (Vol. 37). Berlin: Springer Science & Business Media.
61. Drepper, U. (2007). *What every programmer should know about memory* (Vol. 11). Red Hat, Inc.
62. Hennessy, J. L., & Patterson, D. A. (2011). *Computer architecture: A quantitative approach*. Amsterdam: Elsevier.
63. Manga, M., & Stone, H. A. (1993). Buoyancy-driven interactions between two deformable viscous drops. *Journal of Fluid Mechanics*, *256*, 647–683.
64. Behnel, S., Bradshaw, R., Citro, C., Dalcin, L., Seljebotn, D. S., & Smith, K. W. (2011). Cython: The best of both Worlds. *Computing in Science & Engineering*, *13*(2), 31–39. doi:10.1109/MCSE.2010.118.
65. Drescher, K., Leptos, K. C., Tuval, I., Ishikawa, T., Pedley, T. J., & Goldstein, R. E. (2009). Dancing Volvox: Hydrodynamic bound states of swimming Algae. *Physical Review Letters*, *102*(16), 168101–5.
66. Barnes, J., & Hut, P. (1986). A hierarchical O(N log N) force-calculation algorithm. *Nature*, *324*, 446–449.
67. Brackbill, J. U., Kothe, D. B., & Ruppel, H. M. (1988). FLIP: A low-dissipation, particle-in-cell method for fluid flow. *Computer Physics Communications*, *48*(1), 25–38.
68. Brackbill, J. U., Kothe, D. B., & Zemach, C. (1992). A continuum method for modeling surface tension. *Journal of Computational Physics*, *100*(2), 335–354.
69. Dalcin, L. (2016). MPI for Python. https://media.readthedocs.org/pdf/mpi4py/latest/mpi4py.pdf
70. Enright, D., Fedkiw, R., Ferziger, J., & Mitchell, I. (2002). A hybrid particle level set method for improved interface capturing. *Journal of Computational Physics*, *183*(1), 83–116.
71. Greengard, L., & Rokhlin, V. (1987). A fast algorithm for particle simulations. *Journal of Computational Physics A*, *73*, 325–348.
72. Hunter, J. D. (2007). Matplotlib: A 2D graphics environment. *Computing in Science & Engineering*, *9*(3), 90–95.
73. Harrison, C., Krishnan, H. (2012). Python?s role in VisIt. In *Proceedings of the eleventh annual Scientific Computing with Python Conference* (SciPy 2012).
74. Kaus, B. J., Mhlhaus, H., & May, D. A. (2010). A stabilization algorithm for geodynamic numerical simulations with a free surface. *Physics of the Earth and Planetary Interiors*, *181*(1), 12–20.

75. Keller, T., May, D. A., & Kaus, B. J. (2013). Numerical modelling of magma dynamics coupled to tectonic deformation of lithosphere and crust. *Geophysical Journal International, 195*(3), 1406–1442.
76. Moresi, L., Dufour, F., & Muhlhaus, H. B. (2003). A Lagrangian integration point finite element method for large deformation modeling of viscoelastic geomaterials. *Journal of Computational Physics, 184*(2), 476–497.
77. Moresi, L. N., & Solomatov, V. S. (1995). Numerical investigations of 2D convection with extremely large viscosity variations. *Phys. Fluids*, *7*, 2154–2162.
78. Warren, M. S., & Salmon, J. K. (1993). A parallel hashed Oct-Tree N-body algorithm. *Supercomputing*, *1993*, 12–21.
79. Rider, W. J., & Kothe, D. B. (1995). Stretching and tearing interface tracking methods. *AIAA Paper*, *95*, 1–11.
80. Ricard, Y. (2007). Physics of mantle convection. In G. Schubert & D. Bercovici (Eds.), *Treatise on geophysics*. Amsterdam: Elsevier.
81. Samuel, H., & Evonuk, M. (2010). Modeling advection in geophysical flows with particle level sets. *Geochemistry, Geophysics, Geosystems.* doi:10.1029/2010GC003081.
82. Samuel, H. (2012). A re-evaluation of metal diapir breakup and equilibration in terrestrial magma oceans. *Earth and Planetary Science Letters*, *313*, 105–114.
83. Koumoutsakos, P. (2005). Multiscale flow simulations using particles. *Annual Review of Fluid Mechanics*, *37*, 457–487.
84. Zhong, S., & Gurnis, M. (1995). Mantle convection with plates and mobile, faulted plate margins. *Science*, *267*, 838–843.
85. Zhong, S., McNamara, A., Tan, E., Moresi, L., & Gurnis, M. (2008). A benchmark study on mantle convection in a 3?D spherical shell using CitcomS. *Geochemistry, Geophysics, Geosystems*, *9*(10),
86. Zhong, S. J., Yuen, D. A., & Moresi, L. N. (2007). Numerical methods in mantle convection. In G. Schubert & D. Bercovici (Eds.), *Treatise in Geophysics* (pp. 227–252). Amsterdam: Elsevier.
87. Tornberg, A. K., & Greengard, L. (2007). A fast multipole method for the three dimensional stokes equations. *Journal of Computational Physics*.
88. Dahlen, F. A., & Tromp, J. (1998). *Theoretical global seismology*. Princeton: Princeton University Press.
89. Dziewonski, A. M., & Anderson, D. L. (1981). Preliminary reference earth model. *Physics of the Earth and Planetary Interiors*, *25*, 297–356.
90. Alberty, J., Carstensen, C., & Funken, S. A. (1999). Remarks around 50 lines of Matlab: Short finite element implementation. *Numerical Algorithms*, *20*(2–3), 117–137.
91. Neuenschwander, D. E. (2005). Albert Einsteins Dissertation. Elegant Connections in Physics, http://www.sigmapisigma.org/
92. Caricchi, L., Burlini, L., Ulmer, P., Gerya, T., Vassalli, M., & Papale, P. (2007). Non-Newtonian rheology of crystal-bearing magmas and implications for magma ascent dynamics. *Earth and Planetary Science Letters*, *264*, 402–419.
93. Cooper, C. D., & Barba, L. A. (2016). Poisson? Boltzmann model for protein? surface electrostatic interactions and grid-convergence study using the PyGBe code. *Computer Physics Communications*, *202*, 23–32.
94. Drescher, K., Leptos, K. C., Tuval, I., Ishikawa, T., Pedley, T. J., & Goldstein, R. E. (2009). Dancing Volvox: Hydrodynamic bound states of swimming Algae. *Physical Review Letters*, *102*(16), 168101–168105.
95. Fornberg, B. (1996). *A practical guide to pseudospectral methods*. Cambridge: Cambridge University Press.
96. Morra, G., Chatelain, P., Tackley, P., & Koumoutzakos, P. (2008). Earth sphericity effects on subduction morphology. *Acta Geotechnica*, *4*(2), 95–105.
97. Chatelain, P., Curioni, A., Bergdorf, M., Rossinelli, D., Andreoni, W., & Koumoutsakos, P. (2008). Billion vortex particle direct numerical simulations of aircraft wakes. *Computer Methods in Applied Mechanics and Engineering*, *197*(13–16), 1296–1304.

98. Langtangen, H. P., Mardal, K. A., & Winther, R. (2002). Numerical methods for incompressible viscous flow. *Advances in Water Resources, 25*(8), 1125–1146.
99. Langtangen, H. P. (2006). *Python scripting for computational science*. Berlin: Springer.
100. Lenardic, A., & Kaula, W. M. (1993). A numerical treatment of geodynamic viscous flow problems involving the advection of material interfaces. *Journal of Geophysical Research, 98*, 8243–8260.
101. King, S. D., Raefsky, D. A., & Hager, B. H. (1990). ConMan: vectorizing a finite element code for incompressible two-dimensional convection in the Earth?s mantle. *Physics of the Earth and Planetary Interiors, 59*, 195–207.
102. Mahoney, M. S. (1994). *The mathematical career of Pierre de Fermat* (2nd ed.). Princeton: Princeton University Press.
103. Spiegelman, M. (2004), Myths and Methods in Modeling, Columbia University Course Lecture Notes, available online at http://www.ldeo.columbia.edu/?mspieg/mmm/course.pdf.
104. Korenaga, J. (2008). Urey ratio and the structure and evolution of Earth's mantle. *Reviews of Geophysics, 46*, doi:10.1029/2007RG000241.
105. Smolarkiewicz, P. K. (1983). A simple positive definite advection scheme with small implicit diffusion. *Monthly Weather Review, 111*(3), 479–486.
106. Suckale, J., Nave, J.-C., & Hager, B. H. (2010). It takes three to tango 1: Simulating buoyancy driven flow in the presence of large viscosity contrasts. *Journal of Geophysical Research, in press*. doi:10.1029/2009JB006916.
107. Perez, F. In Memoriam, John D. Hunter III: 1968-2012. http://blog.fperez.org/2013/07/in-memoriam-john-d-hunter-iii-1968-2012.html.
108. Prez, Fernando, & Granger, Brian E. (2007). IPython: A system for interactive scientific computing. *Computing in Science and Engineering, 9*(3), 21–29. doi:10.1109/MCSE.2007.53.
109. Shewchuk, J. R. (1994), An introduction to the conjugate gradient method without the agonizing pain. http://www.cs.cmu.edu/?quake-papers/painless-conjugate-gradient.pdf
110. van Rossum, G. (1998). *Glue it all together with python*. In Workshop on Compositional Software Architectures, Workshop report, Monterey, California.
111. Abbott, B. P., Abbott, R., Abbott, T. D., Abernathy, M. R., Acernese, F., Ackley, K., et al. (2016). Observation of gravitational waves from a binary black hole merger. *Physical Review Letters, 116*(6), 061102.
112. Quevedo, L., Morra, G., & Mller, R. D. (2012). Global paleo-lithospheric models for geodynamical analysis of plate reconstructions. *Physics of the Earth and Planetary Interiors, 212*, 106–113.
113. Morra, G., Seton, M., Quevedo, L., & Mller, R. D. (2013). Organization of the tectonic plates in the last 200Myr. *Earth and Planetary Science Letters, 373*, 93–101.
114. Lithgow-Bertelloni, C., & Richards, M. (1998). The dynamics of Cenozoic and Mesozoic plate motions. *Reviews of Geophysics, 36*, 27–78. doi:10.1029/97RG02282.
115. Lithgow-Bertelloni, C., & Guynn, J. H. (2004). Origin of the lithospheric stress field. *Journal of Geophysical Research, 109*, B01408. doi:10.1029/2003JB002467.
116. Sparks, R. S. J. (1978). The dynamics of bubble formation and growth in magmas: A review and analysis. *Journal of Volcanology and Geothermal Research, 3*, 1–37.
117. Jaupart, C., & Vergniolle, S. (1988). Laboratory models of Hawaiian and strombolian eruptions. *Nature, 331*, 58–60.
118. Allard, P., Carbonelle, J., Metrich, N., Loyer, H., & Zettwoog, P. (1994). Sulphur output and magma degassing budget of Stromboli Volcano. *Nature, 368*, 326–330.
119. Petford, N. (2009). Which effective viscosity? *Mineralogical Magazine, 73*(2), 167–191.
120. Houghton, B. F., & Wilson, C. J. N. (1989). A vesicularity index for pyroclastic deposits. *Bulletin of Volcanology, 51*, 451–462.
121. Hurwitz, S., & Navon, O. (1994). Bubble nucleation in rhyolitic melts: Experiments at high pressure, temperature, and water content. *Earth and Planetary Science Letters, 122*, 267–280.
122. Manga, M., & Stone, H. A. (1993). Buoyancy-driven interactions between two deformable viscous drops. *Journal of Fluid Mechanics, 256*, 647–683.

123. Morra, G., Yuen, D. A., Lee, S. M., & Zhang, S. (2015). Source of the cenozoic volcanism in central Asia. *Subduction Dynamics: From Mantle Flow to Mega Disasters*, *211*, 97.
124. Morra, G., Quevedo, L., Yuen, D. A., & Chatelain, P. (2011). Ascent of bubbles in magma conduits using boundary elements and particles. *Procedia Computer Science*, *4*, 1554–1562.
125. Sumita, I., & Manga, M. (2008). Suspension rheology under oscillatory shear and its geophysical implications. *Earth and Planetary Science Letters*, *269*, 468–477.
126. Kerr, R. C., Meriaux, C., & Lister, J. R. (2008). Effect of thermal diffusion on the stability of strongly tilted mantle plume tails. *Journal of Geophysical Research*, *113*, B09401. doi:10.1029/2007JB005510.
127. Drescher, K., Leptos, K. C., Tuval, I., Ishikawa, T., Pedley, T. J., & Goldstein, R. E. (2009). Dancing volvox: Hydrodynamic bound states of swimming algae. *Physical Review Letters*, *102*(16), 168101.
128. Gargantini, I. (1982). An effective way to represent quadtrees. *Communications of the ACM*, *25*(12), 905–910.
129. Anderson, D. L. (2002). How many plates? *Geology*, *30*(5), 411–414. doi:10.1130/0091-7613.
130. Ballmer, M. D., van Hunen, J., Ito, G., Tackley, P. J., & Bianco, T. A. (2007). Non-hotspot volcano chains originating from small-scale sublithospheric convection. *Geophysical Research Letters*, *34*, L23310. doi:10.1029/2007GL031636.
131. Becker, T. W. (2006). On the effect of temperature and strain- rate dependent viscosity on global mantle flow, net rotation, and plate-driving forces. *Geophysical Journal International*, *167*(2), 943–957.
132. Becker, T. W., Faccenna, C., ÒConnell, R. J., & Giardini, D., (1999). The development of slabs in the upper mantle: Insights from numerical and laboratory experiments. *Journal of Geophysical Research*, *104*(B7), 15207–15226.
133. Becker, T. W., & R. J. O?Connell,. (2001). Predicting plate velocities with mantle circulation models. *Geochemistry Geophysics Geosystems*, *2*(12), 1060. doi:10.1029/2001GC000171.
134. Bellahsen, N., Faccenna, C., & Funiciello, F. (2005). Dynamics of subduction and plate motion in laboratory experiments: Insights into the plate tectonics behavior of the Earth. *Journal of Geophysical Research*, *110*, B01401. doi:10.1029/2004JB002999.
135. Bercovici, D. (1998). Generation of plate tectonics from litho- sphere-mantle flow and void-volatile self-lubrication, Earth planet. *Science Letters*, *154*, 139–151.
136. Buffett, B. A., & Heuret, A. (2011). Curvature of subducted lithosphere from earthquake locations in the Wadati-Benioff zone. *Geochemistry Geophysics Geosystems*, *12*(Q06010), 2011G. doi:10.1029/C003570.
137. Buffett, B. A., & Rowley, D. B. (2006). Plate bending at subduction zones: Consequences for the direction of plate motions. *Earth planet. Science Letters*, *245*(1–2), 359–364. doi:10.1016/j.epsl.2006.03.011.
138. Cammarano, F., Marquardt, H., Speziale, S., & Tackley, P. J. (2010). Role of iron-spin transition in ferropericlase on seismic interpretation: A broad thermochemical transition in the mid mantle? *Geophysical Research Letters*, *37*(L03308), 2009G. doi:10.1029/L041583.
139. Capitanio, F. A., Morra, G., & Goes, S. (2007). Dynamic models of downgoing plate-buoyancy driven subduction: Subduction motions and energy dissipation. Earth planet. *Science Letters*, *262*(1–2), 284–297. doi:10.1016/j.epsl.2007.07.039.
140. Capitanio, F. A., Morra, G., & Goes, S. (2009). Dynamics of plate bending at the trench and slab-plate coupling. *Geochemistry Geophysics Geosystems*, *10*, Q04002. doi:10.1029/2008GC002348.
141. Capitanio, F. A., Stegman, D. R., Moresi, L. N., & Sharples, W. (2010). Upper plate controls on deep subduction, trench migrations and deformations at convergent margins. *Tectonophysics*, *483*(1–2), 80–92. doi:10.1016/j.tecto.2009.08.020.
142. Conrad, C. P., & Hager, B. H. (1999). Effects of plate bending and fault strength at subduction zones on plate dynamics. *Journal of Geophysical Research*, *104*, 17551–17571. doi:10.1029/1999JB900149.

143. Conrad, C. P., & Hager, B. H. (2001). Mantle convection with strong subduction zones. *Geophysical Journal International*, *144*(2), 271–288.
144. Conrad, C. P., & Lithgow-Bertelloni, C. (2002). How mantle slabs drive plate tectonics. *Science*, *298*(5591), 207–209. doi:10.1126/science.1074161.
145. Di Giuseppe, E., van Hunen, J., Funiciello, F., Faccenna, C., & Giardini, D. (2008). Slab stiffness control of trench motion: Insights from numerical models. *Geochemistry, Geophysics, Geosystems*, *9*, Q02014. doi:10.1029/2007GC001776.
146. Faccenna, C., Becker, T. W., Lucente, F. P., Jolivet, L., & Rossetti, F. (2001). History of subduction and back-arc extension in the Central Mediterranean. *Geophysical Journal International*, *145*(3), 809–820. doi:10.1046/j.0956-540x.2001.01435.x.
147. Forsyth, D., & Uyeda, S. (1975). On the relative importance of the driving forces of plate motion. *Geophysical Journal of the Royal Astronomical Society*, *43*, 163–200.
148. Forsyth, D. W., Harmon, N., Scheirer, D. S., & Duncan, R. A. (2006). Distribution of recent volcanism and the morphology of seamounts and ridges in the GLIMPSE study area: Implications for the lithospheric cracking hypothesis for the origin of intraplate, non-hot spot volcanic chains. *Journal of Geophysical Research*, *111*, B11407. doi:10.1029/2005JB004075.
149. Funiciello, F., Faccenna, C., Giardini, D., & Regenauer-Lieb, K. (2003a). Dynamics of retreating slabs: 2. Insights from three-dimensional laboratory experiments. *Journal of Geophysical Research*, *108*(B4), 2207. doi:10.1029/2001JB000896.
150. Gerya, T. V., Connolly, J. A. D., & Yuen, D. A. (2008). Why is terrestrial subduction one-sided? *Geology*, *36*(1), 43–46. doi:10.1130/G24060A.1.
151. Goes, S., Capitanio, F. A., & Morra, G. (2008). Evidence of lower-mantle slab penetration phases in plate motions. *Nature*, *451*(7181), 981–984. doi:10.1038/nature06691.
152. Hager, B. H. (1984). Subducted slabs and the geoid: Constraints on mantle rheology and flow. *Journal of Geophysical Research*, *89*(B7), 6003–6015. doi:10.1029/JB089iB07p06003.
153. Jadamec, M. A., & Billen, M. I. (2010). Reconciling surface plate motions with rapid three-dimensional mantle flow around a slab edge. *Nature*, *465*(7296), 338–341. doi:10.1038/nature09053.
154. Morra, G., Regenauer-Lieb, K., & Giardini, D. (2006). Curvature of oceanic arcs. *Geology*, *34*(10), 877–880.
155. King, S. D., Lowman, J. P., & Gable, C. W. (2002). Episodic tectonic plate reorganizations driven by mantle convection, Earth planet. *Science Letters*, *203*, 83–91. doi:10.1016/S0012-821X(02)00852-X.
156. Quevedo, L., Morra, G., & Muller, R. D. (2010). Parallel fast multipole boundary element method for crustal dynamics. In IOP Conference Series: Materials Science and Engineering (Vol. 10, No. 1, p. 012012). IOP Publishing.
157. Lithgow-Bertelloni, C., & Guynn, J. H. (2004). Origin of the lithospheric stress field. *Journal of Geophysical Research*, *109*, B01408. doi:10.1029/2003JB002467.
158. Stadler, G., Gurnis, M., Burstedde, C., Wilcox, L. C., Alisic, L., & Ghattas, O. (2010). The dynamics of plate tectonics and mantle flow: From local to global scales. *Science*, *329*(5995), 1033–1038. doi:10.1126/science.1191223.
159. Veevers, J. J. (2000). Change of tectono-stratigraphic regime in the Australian plate during the 99 Ma (mid-Cretaceous) and 43 Ma (mid-Eocene) swerves of the Pacific. *Geology*, *28*(1), 47–50. doi:10.1130/0091-7613.
160. Wu, B., Conrad, C. P., Heuret, A., Lithgow-Bertelloni, C., & Lallemand, S. (2008). Reconciling strong slab pull and weak plate bending: The plate motion constraint on the strength of mantle slabs. Earth planet. *Science Letters*, *272*(1–2), 412–421. doi:10.1016/j.epsl.2008.05.009.
161. Zhong, S., & Gurnis, M. (1995). Towards a realistic simulation of plate margins in mantle convection. *Geophysical Research Letters*, *22*, 981–984. doi:10.1029/95GL00782.
162. Stegman, D. R., Schellart, W. P., & Freeman, J. (2010). Competing influences of plate width and far-field boundary conditions on trench migration and morphology of subducted slabs in the upper mantle. *Tectonophysics*, *483*(1–2), 46–57. doi:10.1016/j.tecto.2009.08.026.

Index

G. Morra, *Pythonic Geodynamics*, Lecture Notes in Earth System Sciences,
DOI 10.1007/978-3-319-55682-6

Printed in the United States
By Bookmasters